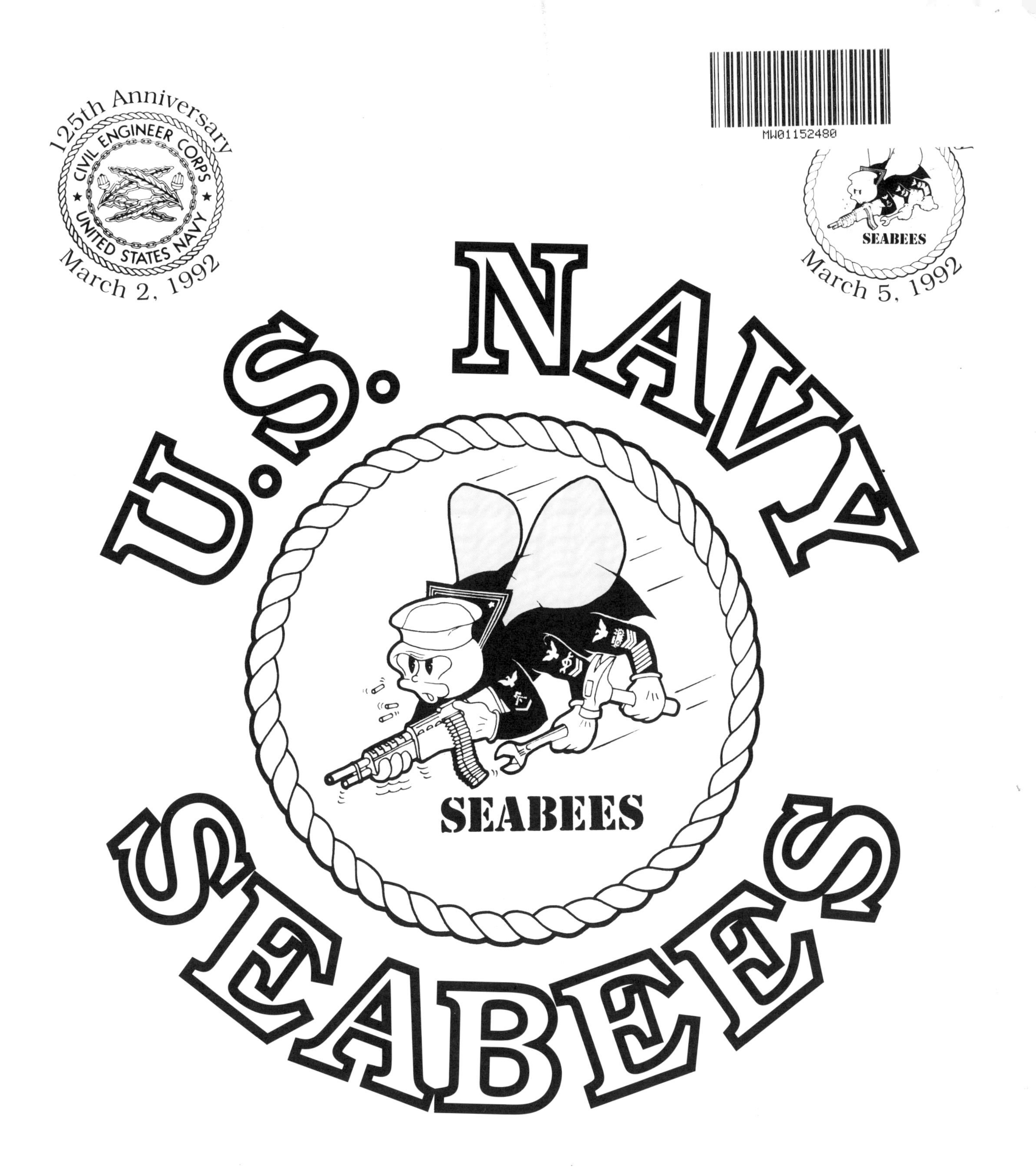

SINCE PEARL HARBOR

JAY KIMMEL

U.S. NAVY SEABEES:

Since Pearl Harbor

Library of Congress Catalog Card Number: 94-069762

ISBN: 0-942893-03-4

Author's Background:

Jay Kimmel, B.A., M.S., Entrepreneur and Writer. Former Certified Rehabilitation Counselor (Eighteen years public and private experience). Previous publications include *Real Estate Investment* (Cornerstone/Simon & Schuster, 1980), *Money Strategy,* (Conifer Publishing, 1982) and CoryStevens Publishing, Inc. as follows: *Savage & Stevens Arms* (1997), *Custer, Cody & the Last Indian Wars* (1994) & *Home Work: Starting a Small Business At Home* (1996).

To order: Contact

CORYSTEVENS PUBLISHING, INC.
640 N.E. 148th Ave.
Portland, OR 97230
(503) 252-9339 • FAX (503) 252-6111 • e-mail: corystvn@europa.com

INTRODUCTION

> *The admiral just dropped around*
> *to chat the other night.*
> *Said he, 'Now boys you're here to work,*
> *but you've been trained to fight.*
> *So if there's any trouble, don't stop*
> *to put on your jeans . . .*
> *Just drop your tools and grab your guns*
> *—and protect those poor Marines!'*
> —Old Seabee Song

The famous "CAN DO!" reputation of the U.S. NAVY SEABEES was first earned for doing essential construction work in the face of enemy fire on the fiercely held island of Guadalcanal (1942). That same reputation for building and defending themselves as needed in times of war or natural disaster throughout the globe has continued to this date.

SEABEES, in close coordination with the U.S. Marine Corps, have served as builders and fighters with distinction throughout World War Two, the Korean War, the Vietnam War and, more recently, the Gulf War (Desert Storm). The Seabees are likewise remembered for providing relief efforts following natural disasters, such as the Mt. Pinatubo eruption, the Chilean earthquake, and civic projects throughout the world (sometimes known as the Navy's "Peace Corps"). It's that spirit of quickly doing what has to be done and of working with whatever is available that distinguishes the U.S. Navy Seabees.

In direct contrast, Viking war parties could once row themselves onto a beach, do what they came to do and row off again. Modern military forces, however, are typically dependent upon millions of tons of equipment, supplies and personnel that can be rapidly and effectively deployed and sustained. Aircraft carriers, for instance, cannot dock at a sandy beach and fighter aircraft cannot land on a pile of rocks or in craggy ravines. Modern militaries depend on *infrastructure*—the ability to quickly establish docking facilities, landing facilities, living facilities, roads, bridges, storage buildings, hospitals, and to quickly move the millions of tons of supplies and equipment as needed under conditions that may be anything but ideal. That's where the fighting SEABEES—builders who can fight and fighters who can build—have proven their worth.

During World War Two this Seabee tradition was described by Rear Admiral O. O. "Scrappy" Kessing with the following: "They're a rough, tough, loyal, efficient bunch of men who don't give a damn for anything but doing the job and getting the war over." General Douglas MacArthur said, "The only trouble with the Seabees is that we don't have enough of them."

In the 1990's there are little more than 10,000 active duty Seabees. However, before the end of World War Two the number of Seabees swelled to more than 325,000. With fierce determination they constructed causeways, docking facilities, roads, runways, hospitals, shelters, bridges, water systems, sanitation facilities, housing and whatever was critically needed at the time.

"It was in the Pacific Theater of Operations, even more so than the Atlantic, that the Seabees would innovate, build, and distinguish themselves most of all during the Second World War. Eighty percent of America's total naval construction force would be assembled along the three vital roads to victory in the North, Southwest and Central Pacific. There, spread across four continents, millions of square miles of ocean, and over 300 islands, the Seabees literally would build and fight their way to victory. In unprecedented feats of wartime construction just part of what the Seabees built reads like Ripley's 'Believe It or Not': 111 major airstrips, 441 piers, over 2,500 ammunition magazines, hospital capacity for 70,000 patients, 700 square blocks of warehouses, housing for a million and a half troops, and storage tanks holding one hundred million gallons of gasoline! That incredible amount of dangerous construction work—and the combat that often went with it—cost the Seabees over 300 lives, and earned them more than two thousand Purple Hearts. Vice Admiral W. L. Calhoun, USN, summed it up most succinctly when he remarked, 'The Navy will remember this war by its Seabees.' " (Millet, Jeffrey R., *The Navy Seabees: The First Fifty Years,* Taylor Publishing Co., Dallas, TX).

The U.S. NAVY SEABEES evolved quickly and in direct response to urgent, wartime needs of a reluctant and grossly unprepared nation. In 1940 and 1941 much of the world was in a state of war. The United States' military at the time of the attack on Pearl Harbor, however, was small, mostly obsolete, and completely unprepared to confront the militant nations of Japan, Germany and Italy (Axis Alliance).

A military void and extreme political sentiment of isolationism gripped the nation after the profoundly tragic sacrifices of the World War—"The War to End All Wars!" The situation was compounded in the extreme by one of the world's longest and most severe economic depressions and the last vestiges of colonialism by the radically military governments of Japan in the eastern sphere and Germany-Italy in the west.

Soon, the world was again at war. Large regions had been at war since the Japanese invasion of Manchuria (1931) and the German invasion of Poland (1939). The lack of U.S. preparedness for international conflict by the time of the attack on Pearl Harbor virtually could not have been more extreme. The U.S. military was supplied with little more than aging ammunition from the World War (including left-over K-rations, and Doughboy weaponry).

Technological advances in the single generation from 1918 to 1941 were profound and mostly favored the militant, Axis nations. The outcome of World War II was distinctly uncertain for the first two to three years at least. Western civilization outside of the Western Hemisphere was at risk of being crushed by fanatic, police-state mentality. There were many tragic instances of loss of lives due to the "new world order" but it was not until the severe shock and sense of betrayal of the attack on Pearl Harbor that the citizens of the United States rallied to the defense of European countries in particular ("western civilization") and fought against fascist control of major segments of the world.

It was in this boiling inferno (a war that would eventually cost 35,000,000 lives) that the U.S. NAVY SEABEES were formed to meet urgent and desperate construction needs in critical locations throughout the world. Their legendary accomplishments have continued to the present date in both wartime and time of major, international disasters. This photo-history is therefore dedicated to all past and present members of the U.S. Navy Construction Battalions—THE SEABEES.

In October, 1941, Rear Admiral Ben Moreell, Chief of the Navy's Bureau of Yards and Docks initiated the groundwork for a Naval Construction Force. This was in anticipation of the ominous military crises developing on two ocean fronts. Then, almost immediately after the attack on Pearl Harbor, Adm. Moreell was authorized to form construction battalions that would work as integrated teams to meet the construction needs of other military operations.

"Seabees," says Alfred G. Don, National Historian, Navy Seabee Veterans of America, Inc., "are both building fighters and fighting builders. From island hopping of World War II and the cold of Korea, to the steaming jungles of Vietnam, the SEABEES have built entire cities, bulldozed and paved thousands of miles of roadways and flattened numerous airstrips in now long-forgotten places."

"The earliest SEABEES were recruited from the ranks of the civilian construction trades and were placed under the leadership of officers of the Navy's Civil Engineer Corps. With emphasis more on experience and skill than on youth and physical standards, the average age of SEABEES during the early days of the war was mid-thirties, giving rise to a famous Marine Corps barb: 'Don't strike a SEABEE, he just might be a Marine's father.' " (A. G. Don).

New to the war in Vietnam, Lt. Harvey M. Henry, Ninth Seabee Battalion, arrived at the improvised hospital at Da Nang, shortly after completing his one year medical internship (June, 1966). He was met by rocket fire that all but demolished the structure and produced 88 casualties and two fatalities. Dr. Henry persevered with the assistance of two corpsmen in treating the wounded under the most adverse of medical situations. For his extraordinary coolness and skill in saving lives he earned the lasting respect of the Ninth Mobile Construction Battalion *and* the Navy Commendation Medal. The Seabees, in turn, reconstructed a 200-bed hospital to treat battle casualties and Vietnamese villagers as possible.

In another factual episode, a group of Seabees were carving a road through a 2,000-foot peak in Vietnam known as Monkey Mountain because the rugged country was inhabited by large baboons as well as Viet Cong. Temperatures often reached tropical extremes of 130 degrees and weapons were kept close to the tools being used. Beside a rock crusher someone had fashioned a crude sign that read: "Your tax dollars at work. This road built by the Seabees for the convenience and comfort of the United States Marines."

Out of the tropical heat a Huey helicopter swooped down and, from the swirl of dust, Lt. Gen. Victor H. Krulak, commander of the Fleet Marine Force, Pacific, started walking toward the first available Seabee. The general had come to inspect the Hawk anti-aircraft batteries on the mountain and to check on the road's progress. His comment to a Seabee at the edge of the perimeter was, "How do you tell these Seabees from the baboons?"

"No problem, sir," said the young Seabee. "The Seabees are smoking cigars."

The general glanced around the perimeter and noticed that every Seabee had a cigar in his mouth. Smiling, the general climbed aboard the Huey and was gone.

Special thanks are extended to the following persons for their essential contributions and often on-going assistance to make this book possible: A. Clark Fay, Louis Schroen, Henry B. Lent, Jeffrey R. Millet, Dr. Vincent A. Transano, NAVFAC Historian, Port Hueneme, CA, Alfred G. Don, National Historian, Seabee Veterans of America, Inc., Elaine McNeil, Public Affairs Officer, Alexandria, VA, U.S. Navy Seabees Memorial Museum, Gulfport, MS, National Archives, USN, Bureau of Yards and Docks, Washington, D.C., Mrs. K. and others named within.

Early SEABEE Poster*

ENLIST WITH THE "SEABEES"

October, 1942

N.A.B. "Strawstack" South Pacific

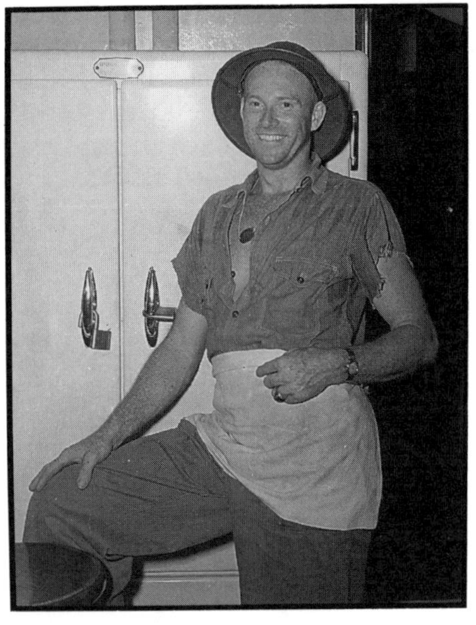

A. Clark Fay, CM 1/C "Seabees"
U.S.N.R. C.E.C
Aboard the Seaplane Tender *USS Tangier*

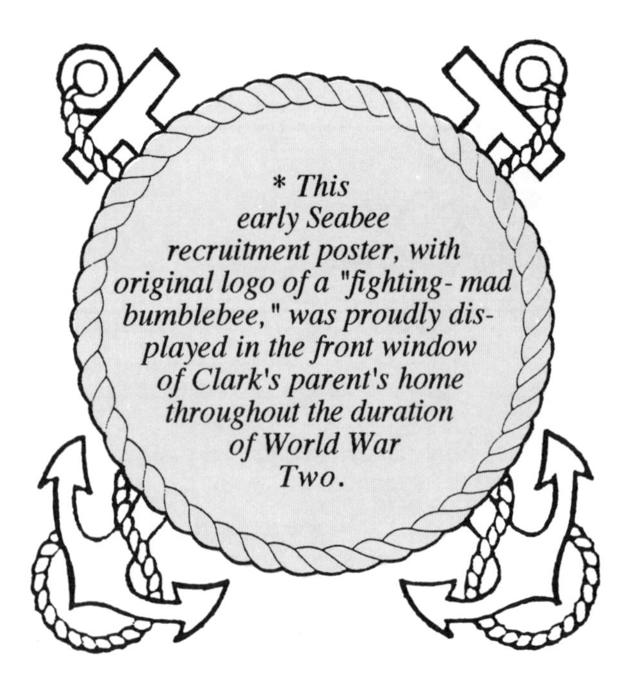

* This early Seabee recruitment poster, with original logo of a "fighting- mad bumblebee," was proudly displayed in the front window of Clark's parent's home throughout the duration of World War Two.

This book is dedicated to all U.S. Navy Seabees, including A. Clark "Red" Fay April 18, 1910 to February 11, 1997

SEABEES IN THE SOUTH PACIFIC, 1942-43

A. CLARK "RED" FAY

REMEMBER: NO MAIL FOR SIX WEEKS, the land crabs, TOM CAT'S CREW, "Shoutin'" Scouten, Oregon's short circuit, No beer, USO shows in the rain and mud, July 19, 1942, 1,008 miles out of San Francisco, October, 1943, "Dreaming of a White Christmas," SAN FRANCISCO, HERE WE COME!

The following is a short story about a few men, each trying to do his part toward a better land in which to live. Little knowing the "bad time" in store for him. Little knowing that he was to live in and around tropic fever and then to overcome and live a short life in a paradise of duty, when compared to the experiences some of his fellow servicemen. This story is dedicated to those men and their families.

PORTLAND, OREGON TO GREAT LAKES NAVAL TRAINING STATION, ILLINOIS
Boot Camp, May, 1942

It all began when we volunteers were notified to report to the railroad station in downtown Portland for transfer from civilian life to life in the service of our country. Destination: the Great Lakes Naval Training Station. This training base had been in service during World War One. Here thousands of men were taught the ways of the Navy.

On the dock in Portland were thirty, very diverse men who volunteered for duty in the newly organized branch of the Navy known as the Construction Battalion. This nationally organized group was later referred to as "Seabees." The Navy had furnished one chief to oversee the train ride to the training station.

Now, almost fifty-three years later (i.e., 1995), I will share my personal experience as a new Navy man based on notes at the time and what memory will allow.

The trip to the Great Lakes was jammed with excited and sometimes rowdy young men but was otherwise uneventful. Time was filled with card games and walking through the train talking to the other passengers. There was a plentiful supply of liquor. At any stop of fifteen minutes or more, some of the men would go and buy more liquor—a male rite of passage, I guess.

Three very different men and I became buddies. One was a cowboy from Montana, one man was a recently discharged Coast Guard sailor I had known at Astoria [Oregon] the year before. The third man was a Native American from South Dakota.

There was a meat cutter from Eugene [Oregon] whose brother had attended the same high school as I. We were never buddies, just two men who shared the same home town.

Arriving at Great Lakes Naval Training Station, we were marched through high gates with armed guards on both sides, then to a barracks that housed at least two hundred men. Here we were to string up hammocks and learn to sleep in them. This activity was really humorous. It was almost impossible to get in a hammock and not turn over. One of the newly appointed chiefs fell out and broke an arm.

Each day we were marched to a different building for shots. At each meal we marched to and from the mess hall, each platoon trying to out-sing the others. We were being taught military obedience. When the officer in charge said, "Do it!" You DID IT THEN. There was little doubt that life had changed.

For a few days some of us were assigned to the galley (kitchen) as mess cooks (helpers). I drew duty in the gear locker (brooms and mops). This little space opened on the hallway that led to the chief's mess. As one of my buddies went by with a special dish, he would always pass me a plate. The chiefs ate better than the other men. Also, there was a hot shower there and I enjoyed my first hot shower since leaving Portland. Being assigned to the gear locker, I was able to walk around the large galley. The smell of boiling beef in the thirty-gallon pots nearly made me sick. The regular cooks would tell me to leave, as "boots" were not allowed in that area. One night they were having steaks, and it was at that moment that I learned if you want to eat, you must be in the galley. Later, down in the South Pacific, this personal experience paid off.

At one time I was assigned to guard duty. Before the four-hour time was over, I fainted and knocked my head on the cement floor. I learned about the "sick bay" (hospital) from that personal experience. It was not serious, so my stay lasted only two days.

The men at boot camp were supposed to be taught to swim, but that never came about. After four weeks of doing everything on the run, we were paid and allowed to go into the town for a bit of R & R ("rest & relaxation"). I visited Minneapolis, spent much of the time riding horseback at a commercial stable and returned to the base with a little money still in my pocket. Riding and dancing had been among my favorite pursuits while going to college, and I didn't know any dance partners in Minneapolis.

Soon, we were on our way to Norfolk, Virginia, and more training. This trip was more noisy as we traveled by troop train. The train was made up of standard pullman cars and we ate in the dining cars. Some place along the line we were side-tracked for some time. We were in the shadow of a four-story factory. All the windows were filled with women shouting and waving. When in the South Pacific, I wrote to several of the girls on the fourth floor. One answered and we corresponded for several years.

Now being on a new base, the routine was the same, but the officers were different. There was a call for mess cooks. I said to Perko, a buddy from Portland, "Let's take the job." We took the garbage detail. This sounds bad, but we only worked for fifteen minutes after each meal time and the rest of the day was free. Besides, we were where the food was!

Soon, we were aboard another troop train bound for Port Hueneme [Why nee' me], California. This was to be the new embarkation center for the Seabees. I grabbed a four-day pass and flew to Eugene. Some other men on the west coast also drew passes for a few days. The highlight of being stationed here was liberty in Los Angeles. The first night out taught many men a cold lesson—after six o'clock, sunny California turns freezing cold. Many of the men wearing only their dress blues were later caught in 30 degree temperatures, and almost as fast parted from their money.

We were now a full size battalion—twelve hundred men. The supply list was pages long, yet our construction battalion was scheduled to ship out without one jeep or even a dump truck. How much can a small group of men do with picks and shovels only? On July 14, 1942, the word went around that "We ship out tomorrow" to some place called "Straw Stack." And so we began.

SHIPPING OUT

Port Hueneme, CAL., 15 July 1942. Seventh Battalion Seabees moved out for island "K," known as "Straw Stack." I have been in sick bay for the past three days with a fever, cough, and dismal headache. No fever by the third day, but a severe headache kept me in bed.

We were now rolling along the coastline with the Pacific breaking on the nearby beach. There were some beef cattle and some oil wells out in the surf. One Army search light unit was set up. The men were in high spirits and excited. The last view of the U.S. coastline was the historic landmark of "Nob Hill" atop the City of San Francisco. Within a couple days, we were over one thousand miles west of the "City by the Bay." We were aboard the *USS President Monroe* with a total of 1103 men in the crew. After seventy-two hours, we were 1,440 miles

U.S. Navy Seabees in the South Pacific

World War II, Korea, Vietnam

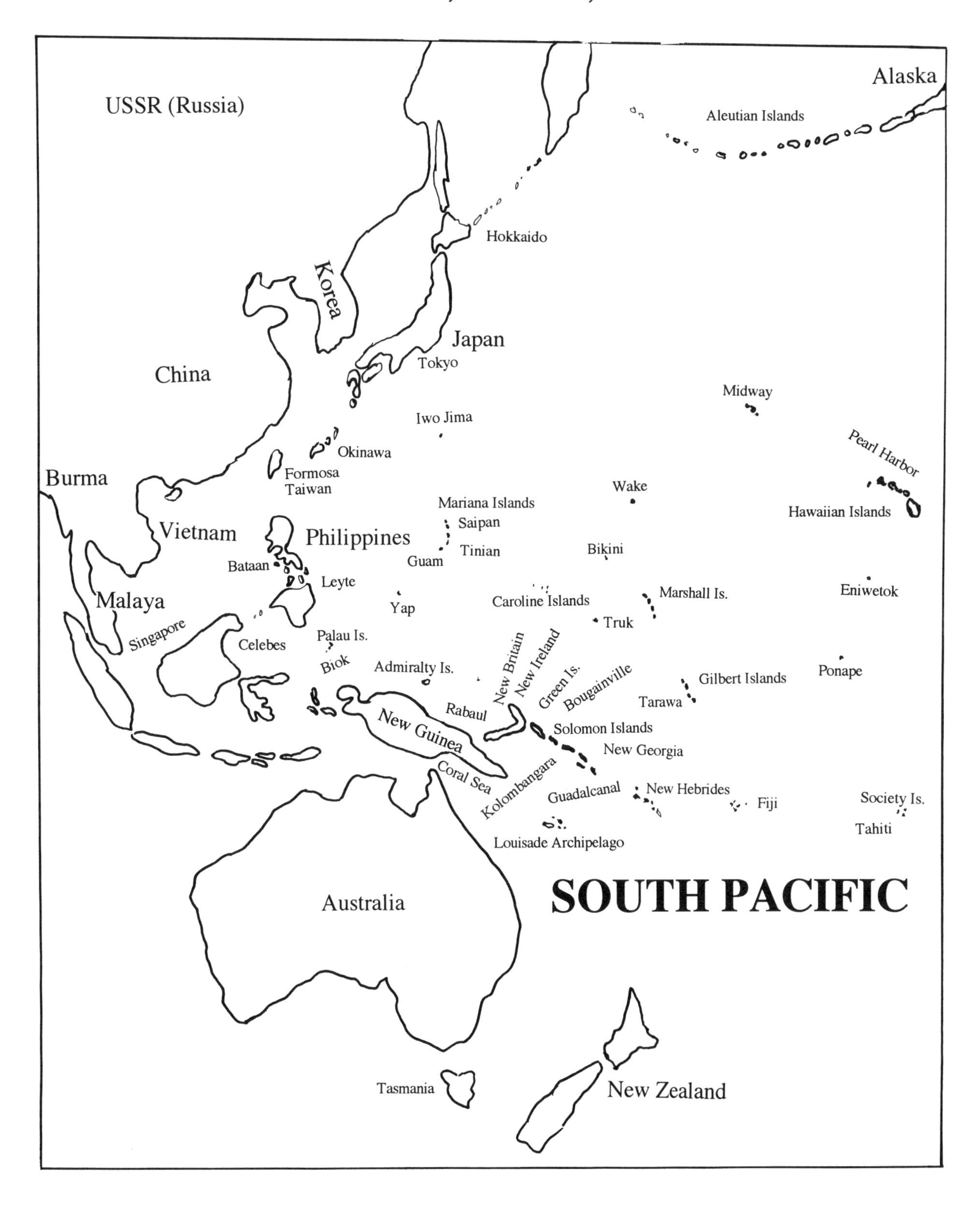

from shore. Very few white caps were visible over this unlimited expense of water.

Eight days out in the Pacific and I served guard duty from 0000-0400. The night before I had slept on the deck. It rained hard and was wet in the morning. It seemed natural enough to take a shower in the rain. The hole (our living quarters) was getting dirtier each day. With some exceptions, the food was unappealing. On the ninth day out, the equator was crossed. The monotony of travel was relieved that day by the King Neptune initiation of the neophytes into the "Equatorial Club." There were lots of pranks such as eggs broken on heads and in shirts, bad haircuts, and dousing with firehoses. Anyone, including officers, found still dry was thrown into a tank of sea water.

Ten days out and the weather was very windy. Some sharks were observed along the starboard side of the ship. The upper gear and the spar poles of the ship created interesting silhouettes against the evening clouds. The moon was very high. In the moonlight, the severe sunburns of the men were apparent and left various patterns (depending on how T-shirts were worn) on their bodies. Boxing and card playing were the most popular daily diversions. Virtually anything that took our minds off the poor quality meals would have been popular.

Eleventh day out. Very warm weather even in undress blues. Men who wanted to use the lounge to write letters were required to wear undress blues. Actually, it was quite difficult to write due to the pitch and roll of the ship. There was a good church service on that day (Sunday) and many were present. Many of the men slept on the top deck wearing undress blues with leggings, coveralls, sweater and watch cap for warmth. The "Southern Cross" was visible in the night sky.

TROPICAL PARADISE

July 30th: We went ashore at Pago Pago. After twelve days aboard a navy transport ship, it was a great relief to arrive at a friendly island in the South Pacific. The sandy beaches and scattered palm trees surrounded impregnable jungles. The image of tropical paradise, however, was in stark contrast with reality. The new recruits were actually undergoing distinct changes from modern living to, as Admiral Halsey said, "the worst of living conditions."

By that time, we were experiencing serious changes. Page after page of censorship regulations were posted on every bulletin board. My letters home were therefore rather short.

Early on the morning of the thirteenth day out, our convoy was circling off-shore a group of islands whose irregular coastline was outlined by the constant wash of the Coral Sea upon the coral reefs. These islands were a very welcome sight after seeing nothing but water for the past twelve days and after watching island peaks appear and then fade into the horizon for the past two days.

As we neared the new islands, sailors came to life under the fast rising, tropical sun. We were greeted by a Marine observation scouting plane which swept the ship almost at mast level. The gesture was returned with the hail of thousands of cheers from the travel weary men. This plane was our first sign that we were in friendly waters. A few minutes later the ship's navigator informed us we were nearing American Samoa. It was the tropical paradise of a sailor's dreams—a land of beautiful Polynesians.

During our continued circling around various islands, we studied its rugged landscape covered by either semi-thick or densely thick jungle growth. Small clearings could be spotted here and there on steep hillsides. Later, we learned that each clearing represented a native family and that each was connected by a narrow jungle trail.

Suddenly, a plane zoomed out of a hidden valley and passed over the crest of a surrounding hill. A second plane came into view just as quickly and then settled out of sight in the vast jungle.

By mid-morning, a harbor pilot came alongside and he started his tedious course through the thickly-laid minefield that protected our American interests below the equator. Nearing the mouth of the world's most "beautiful harbor," we soon took on a writer's mood in describing this tropical paradise.

Seeking the most advantageous viewpoints topside, the men began climbing to the mast tops for quicker and better views of their new home. Being lucky, I reached one mast before a shipmate and was atop that sixty-foot pole in nothing flat.

Weaving a delicate pattern with its wake, our ship would occasionally catch the sun above the millions and millions of bubbles created by the slowly turning prop. On a massive scale, it resembled a small boy playing with his soap bubble pipe in a tub of suds. The image created was a never-to-be-forgotten mental picture of the ever-changing background of coral reefs and white, sandy beaches that were constantly washed by the rolling surf.

Details of men were busy with various jobs. Some men returned from the nearby airstrip with reports of recent bombing. Our job—to stop Tojo's southward march—was becoming more vividly real.

At mess call, we were greeted by a smiling Polynesian woman employed as a waitress in the contractor's mess hall. Meals were served family style and the usual griping was noticeably absent as the conversation turned to laughter and the men enjoyed their first good meal in two weeks.

Here we learned that the beautiful white beaches were once poisonous coral reefs. Swimming in the tropical, blue-green water was dangerous because of the coral reefs and sharks that inhabited the shallow areas. If a bit of coral became embedded under the skin, it could produce almost incurable sores. Advanced cases resulted in the dreaded disease of elephantitus.

TO THE FRONT

A few days later, a fast flying scuttlebuck came out from the dock with orders that we would be moving north. Excitement and speculation increased by the hour. However, it was several days before sailing orders were posted. In the meantime, we enjoyed the islands.

There were 162 of us in one dorm. We had been building double-deck beds. Due to space limitations, they were too close together. I was one of the lucky ones with inner-coiled springs on my bunk—a small comfort the civilian workers no doubt hated to leave behind. The friendly Samoans would also be missed. My shipmate, Onuschuch, was an enterprising businessman. To obtain postcard photos to sell to the men at 25 cents each, he first had to walk about twelve miles to buy them from a native woman who was married to a German spy. The spy, of course, was being held in a prison camp in the U.S. and the postcard sales were perhaps the woman's only means of support.

According to an old chief, a volcano had erupted about 84 miles away and sent lava dust over Samoa. He claimed to have been a small boy at the time. Others were willing to share bits of their language such as, *mo mo* (red), *tofa* (hello), *Laua-laua* (shirt), *Fafi tiga* (thank you).

Arriving in port, we were issued packs. We were traveling with three other ships—a freighter, a passenger ship, and one cruiser. It was uncomfortably warm and the men slept on the deck. The officer in charge of the ship was one rotten fellow—just like "boot camp!" He made a big fuss about "dress of the day," being addressed as "Mister," etc.

Reaching the docks, to our amazement, we saw several more destroyers, or "cans." During the early morning hours, the dawn patrol had been visibly increased and the sight of the additional "cans" put the men's nervous systems on alert.

To the west were the Fiji Islands. They had just been bombed and were badly in need of a second air field. Still further west was New Caledonia. The Japanese were driving hard to reach those little-known islands. Possibly we were headed for the Solomon Islands.

Setting out to sea, our course was concealed by the early morning blackness. Then, with the rising of the morning sun, our course was estimated to be toward the northwest. For days we continued cruising, always west by northwest. Each ship of the convoy kept an exact distance between it and the one ahead while our protective "cans" ran circles around the convoy. This ever-present alert of the "cans" kept the men at ease—

but then came the endless black night. Shadows on the reflecting waves from various ships were our only evidence of companions on those nights of patrolling.

During each day, time was consumed by different tasks aboard ship. One main order was "Don't throw anything over the side." All garbage was ground up and dumped at night in case a Japanese sub should skim the sea for evidence of the convoy's movements. Potentially, the enemy might be able to determine the origin, destination and the number of men aboard the ships by the refuse. Thus, the night's dumping was well scattered by the first daylight. The exception was during the full moon. Those nights were like days. One night, in fact, the moon kept circling our ship. We were biding our time. In the moonlight, we observed the destroyers that had moved in closer to the mother convoy.

During these days, there had been the sudden departures from the set course in a wide fan around a "can" or two tracking down a radar beep: always there was the potential that an unexpected enemy sub could be following. However, these sorties proved negative, and again we relaxed in the tropical sun and strong ocean breezes.

One order of the day was "Wear life preservers at all times. Have a little food in your ditty bag. Nothing heavy." Exercises were described on how to land in the water. These daily memos kept us on our toes while sitting idly by on the slow moving transport.

GUADALCANAL

In the distance, a tropical storm was blowing hard. Little did we realize that the supposed lightning was cannon fire from a sea battle, and the thunderous roars were actually gunfire. Near by, within ear sound, a terrific sea battle was being fought just over the horizon.

Then, as before, the morning sun indicated that our course had been changed. This time we headed due south. Later we learned that the Battle of Coral Sea was being fought at a dear price, and we were only thirty miles off-shore.

For another three or four days, we set our course by the Southern Cross. Soon we were circling a group of palm-covered islands. The silent morning was suddenly broken by the thunderous roars of the huge army B-17's and the deeper roar of the Navy's seaplanes as they flew off to the north. This sudden burst of action occurred at 0330 Navy time. Obviously, our sleep above decks was broken by loud noises and the chill of the early morning. The date was August 12, 1942, and the day began early.

More planes flew away to the north, and a large duck-like Catalina started its endless patrol over the vast expanse of the Coral Sea. These planes could cruise as slowly as one hundred miles per hour and just skim the waves while doing so.

Once again our convoy steamed slowly behind a pilot boat edging its way into a harbor rich in scenery but well protected by its hidden mines. This same location was to be a "Davy Jones Locker" for the ill-fated *USS President Coolidge* a few weeks later.

NEW HEBRIDES

Word from the ship's navigator was that we were making port in the New Hebrides Islands. Curiously, these islands were noted for their cannibals and headhunters—something we'd not been trained to relate to. Visually, the islands presented a mystic picture of mountain peaks reaching into low hanging clouds. Deep valleys ran into the interior and were lost in dense jungles. Coconut groves ran from the shallow beaches back to the dense jungle that was spotted with giant mangrove trees and red-topped huts. Each of the red-top huts represented the homes of white plantation owners. It was like visualizing the meeting point of a very primitive tribe and the beginnings of a very different civilization.

After entering the harbor, which was only a channel between two islands, orders of the day were for the issuance of rifles and packs. We were now a military outfit in enemy territory. Possibly half of the men had even seen a 30.06 rifle before, much less actually fired one. Here we were to make a beachhead—only someone had beaten us to it. The Marines had landed. Yes, we could see Marines on the beaches and in the few clear areas. Coastal defenses, however, were not apparent.

Orders to go ashore at 0600 were received. Because of delays, I went ashore with a working party about noon. Immediately we began working on the dock, a dirt-filled wharf. It had been built by using coconut tree logs as a retaining wall and then back-filling with sand. The fast moving current along with the fast moving "Higgin's boats" kept washing the sand out into the channel.

One of the chiefs asked me, "Do you want to do guard duty instead of being a longshoreman?" I agreed and drew a double watch in the coconut groves. It was a 250-acre grove. The only people seen on the island were small, black pygmies, each carrying a long and mean-looking harvest knife. The strangest sounds came from the coconut trees themselves. All through the 0000-0430 watch, the lower limbs of the coconut trees dropped huge limbs. The swishing noise and thud was something to behold the first few times it happened.

Our first purpose had been accomplished. Safely secured to the huge cleat on the dock and shore, our ship rested easily in the blue-green Pacific water after a safe crossing and success in bringing much-needed supplies to another remote island base.

Immediately, the operating booms began their endless screeching from sunrise to sunrise as they lifted the precious cargo out of the bowels of the ship. Care was taken in using the makeshift dock. Much of the cargo was sorted and then handled by Samoan longshoremen who worked barefooted wearing their customary skirt fastened with wide belts.

Several sailors and marines on shore shouted, "Anyone from Jersey?"

"Who's from Texas?"

"When did you leave the states?"

"Y' bring any beer?"

About a hundred men shouted back, "Yeah, twenty thousand cases!"

Soon the whine of the many winches were drowned out by the shouts of men on aboard and ashore. The high flying cargo nets were hidden by white-clad Seabees making their first liberty in a strange port. This was their first overseas base—since leaving the world's most famed Southwest Pacific Port of Pago Pago.

COCONUTS FOR A QUARTER

Actually, there wasn't much of a town. There were only few wooden buildings, much like any other small coastal town. A deserted stone church reminded me of the story *Hurricane*. There were a few of Henry Ford's earlier models half buried in the coral sand. There were various remnants of the earlier day trade goods that had long since lost any functional value. Sidewalk merchants offered shells, postcards and other novelties for sale. The young women wore grass skirts that bore the noticeable trademark of "Made in the USA." The commercialism and discovery of the origin of the grass skirts quickly erased the romantic glamour of Seventeenth Century images of South Pacific Islanders. Coconuts, for example, were regularly priced at 25 cents each, yet they grew by the thousands only a few feet away and many feet higher up.

During a short hike into the interior, I decided to pick some of the fruit. The effort was not so easy as the photographs on the travel folders suggested. Hand over hand, or more correctly, arms around the trunk and laboring strenuously for some time, I reached the clusters of coconuts that hung under the bending fronds. Once up there, I discovered coconuts won't come off before cutting a strong, fibrous cord that holds each in place. Descending the tree was a very painful task and, much to my sorrow, I lost huge patches of skin from my arms and legs. It made the 25-cent price seem much more within reason. Also available were fresh bananas, pineapples, and many other tropical fruits that were not so familiar, such as papayas and mangoes.

After reaching the ground, I learned that I could have traded a native one cigarette to perform the same task with speed and grace, drop the coconut to the ground and peel the heavy husk and split it without loss of the milk.

SETTLING IN

By mid-afternoon, and after returning to the ship, I found details of Seabees busy sorting gear as several units were moving to their new camp areas. Being assigned to a dock detail, and reporting about one hour later, I enjoyed another tour of the native town. The Navy and Marines had set up a theater, established a ship's service store, and sent out patrols over the island to maintain peace. Only the week before, a sergeant had strayed off limits, and when he failed to return to camp, a search party located his decapitated body back in the bush. A native bushman was thought to be responsible for the crime. It was one of those vivid reminders that an "OFF LIMITS" sign meant exactly what it said.

Stateside contractors were strolling to the docks, each having been overseas for several months. They were soon to be shipped back stateside. Tojo's southward march mandated the return of civilians. The Seabees would finish what they had started, not for conquest, but for defense.

Transport from ship to an already set-up camp was the continued order of the day. Eventually moving into a barracks we found hot and cold showers. A few lucky fellows salvaged some inner-coil mattresses that some fast-departing contractors had abandoned. Bunks were double-decked. By night, two companies were housed inside just as a tropical monsoon drenched the area.

Thoughts sometimes moved to questions about our first line of defense. On shore there were some American-made trucks and assigned military personnel. Also on shore there were bands of black pygmies, each carrying a long knife, who disappeared and reappeared at will. Occasionally there was a question about the immediate enemy being the pygmies and not the distant Japanese. Rumors of cannibalism didn't make it easier to sleep at night.

ISSUE OF FIREARMS

Growing up in the extreme western part of the United States, I had used a small caliber rifle from the time I was very young. I even had memories of my dad helping me hold it to fire at a paper target. The issuance of rifles gave me quite a moral lift at the time. There was just one catch—no ammunition. The men who were unaccustomed to firearms could not be trusted with loaded firearms unless we were attacked. The alternative to securing live ammunition, unbeknownst to the officers in charge, was to seek out men who had some—the Marines. Boy, did they laugh when the word was passed that Seabees had been issued rifles but no ammunition! About half a dozen men did acquire live ammunition on their own.

The Marines were fully prepared to find their humor at the expense of the newly-arrived and untested Seabees. Actually, there were only fifty Marines stationed on the island, and they had a total of five machine guns, not a sufficient defensive force to repel one of Tojo's expeditionary task forces. Within a few days, however, everyone was issued ten rounds, or two clips, of live ammo. An unfortunate incident soon followed. A disgruntled Seabee, who was totally peeved at a chief, decided personally to alter the chief's views about lording authority over men his own age and older. This obsessed character fired several shots at the chief who took refuge in a signal tower. The next day all ammo was recalled, except for those of us who quietly retained our private stock. The incident passed, and the men continued a never-ending stream of moving supplies onto the island.

BUILDING WHARVES

A fast running tide created nightmares for the initial dock-building crews. The men tried again and again to hold steady long enough to make a small, dirt-filled wharf. The first wharves had been washed over by the pounding of heavy waves caused by the action of boats and the constantly running tide. There were moments of tension during construction as the small boats were tossed about on the high waves and could have crashed down on workers along with the salt spray of the whitecap waves.

With increasing water traffic and scarcity of heavy trucks, it was evident that much of the work to be done would be accomplished by hard, manual labor. In spite of best efforts, the continued running of the tide and lack of heavy equipment, the wharf slowly dissolved into the sea, again. However, with the Higgin's boats, this did not slow the landward march and transfer of supplies. The solution was to drop the forward landing ramp and a quick exit was accomplished.

Our reception on the "Good Earth" was highlighted by a group of pygmies who wore little or nothing except for bone accessories in their noses, ears and hair. Each man waved a long knife over his head. Up close, this sight reminded us of the adolescent adventure stories about cannibals and headhunters that inhabited these remote islands. After a few words from the French planter, the pygmies retired to their foul-smelling huts just a few feet inshore from the ever-widening road.

As the day wore on, details of men formed the often-used-train-of-men unloading the boats as the boats fought the fast tide. Occasionally a large block of carts would wash away.

BOMBER ONE

Two days later found the battalion setting up camp in four different locations. My outfit, Company B, was between the jungle, a coconut grove and the sandy beach. Here, we were completely hidden from planes above and passing ships at sea, yet only a few hundred yards from the newly hacked-out air strip to be known as "Bomber One." Here we would see the red blood of our airmen wasted away on the coral sands. Here, also, was the daily news broadcast with reports about action north of us.

"Bomber One" was a rough strip of sand between rows of coconut trees, assorted mangrove trees, and brightly colored green plants covered with the yellow fruit of the papayas.

Soggy, mushy sand patches were soon transformed into a hard surface runway carrying the four engine B-17's on their early morning bomb runs over the hard-pressed Guadalcanal battlefields. Navy torpedo dive-bombers soon rested here after a patrol for Japanese shipping. The huge Catalinas, with combination seaplane and wheel landing gears refuelled prior to each daytime vigil over the endless ocean.

FIGHTER ONE

A few miles inland, Company A was improving "Fighter One" for the Marine flyers. All they had for the light planes were two rows of trees removed and a road grader making passes down the runway every day or two. One small native hut was the only sign of life on this section of the island. For security reasons, the huts were soon removed. This native area was completely hidden because of the brush construction. Even the corner posts of the thatched huts branched out covered with leaves as the native poles took life from the rich jungle soil and much rain.

The coconut grove, owned by a Frenchman, was fast becoming a beehive of men transforming a supply base into an offensive and defensive base. For obvious reasons, the Frenchman and his family had to be relocated to another island. The same action was necessary with the other English and French planters as well. From this base, essential supplies were the lifeline for the First Marine Raiders battling for Henderson Field on the Solomons.

BACK BAY

A little distance away was the second bay of quiet water. Some merchant ships found a safe moorage behind mines and a net. This was a spectacular setting for one of the busiest harbors of the South Pacific. Men aboard ships could fish for and catch sharks in the bay. Later, the navy set up concrete dry docks which had been towed over by sea-going barges from Pearl Harbor.

About the same time there was an unexpected tragedy when a section of dry dock capsized and drowned several sailors. Three months later, our warehouse crew had the displeasure of discovering the corpse of one of these men. The body was placed in a nearby morgue and was visited periodically by different men out of respect and to briefly escape the drudgery of monotonous routines, week after week.

From this small harbor, the quickly unloaded supplies of gasoline, war materials and bombs were soon turned loose on the Japanese in reprisal for Pearl Harbor. To our amusement, the different-sized bombs each carried a different, select message for its intended targets from the defense workers at home.

EARLY LIVING CONDITIONS

Our company, living under primitive conditions, and at times living on food that made cat food sound attractive, soon made Bomber One resemble a modern airfield. Coffee was typically kept hot with an open fire on the beach. Daytime temperatures were always hot and humid. Sleep was possible only because we were dead-tired and insensitive to the occasional monsoon downpours that flooded our tents.

Today several of us had completed an interesting trip to ship and back ashore in a Marine landing barge. The bow lines fouled and a second barge rammed us. It was a close call for me especially, and I felt lucky to survive the experience. The sea had been quite rough that day. Soon everything was back to normal—eating sandwiches, even for breakfast, almost daily since landing, and working non-stop. In particular, I remember not having a bath for six days running and being extremely dirty. The weather remained warm despite being the winter season. The flies were especially noxious. No baths. No water. No light. No head. Lots of dust, sand, and sweat.

Saturday, August 15. I enjoyed a dip in the Coral Sea. I spotted an octopus about fourteen inches across. Later I took a fresh water shower and certainly felt better. Our detail built a head from small trees on this day. Since nails were scarce, it was necessary to braid bark and tie branches together with our primitive rope.

The food at our new campsite was much improved. Actually we had a breakfast of pears, slices of white bread, pineapple preserves, tomato juice and coffee. It also tasted better now because we were working shorter days (during twelve hours of light). Maybe it was also because the smaller quantities to prepare allow the cooks to give attention to actual flavor. Despite the galley being located in a *follie* (native hut) with a dirt floor, no screens and stoves fueled by gasoline!

AIR RAID & WAR NEWS

Yesterday we had an air raid warning. Four prop planes headed this way. One was shot down by ground fire and the others turned back. Later on the same day we traveled to headquarters camp. Everyone there was very busy. The medical department was erecting lots of Quonset huts that were each 16' x 32' with thatched roofs to be less visible from the air. Real excitement for the day was confined to face to face contact with several spiders that measured six inches from the end of one leg to the other. Their webs were quite strong considering the fineness of the fiber. The island was also infested with green lizards.

Back at camp, as I sat on the edge of my cot, I had an unobstructed view all the way to the eastern horizon. The blue-green color of the calm Pacific was magnificent. The ocean spreads out on a coral reef approximately 100 yards and makes a melodic humming noise except when it is disturbed by a strong gust of wind. The sound then transfers to the rustle of over-sized tropical leaves that remind me of Oregon laurel, only much larger. There were ten men to a sixteen-foot-square tent and since we were overcrowded, I often give up my cot in favor of a hammock strung between two nearby trees.

News is scarce, but various reports brought in by the bomber pilots indicated that the first landing party of Marines were slaughtered by the Japanese. Some men were tied to trees and bayoneted; others were decapitated. One pilot reported that the stench was unbearable and the battlefield abandoned. He was able to return with two .25 caliber Japanese rifles and sundry articles of war.

At night some men attended shows given either at the Marine or Army camp. Personally, I usually hung around camp. The previous night we had a lunch in the galley. Pop Nelson, Jack Flynn, Roworry, Schroff and a couple of other Portland men were there. The relaxed time together passed quickly.

Rumors of the island being inhabited by head-hunters were common, and all men were advised to carry rifles and to stay out of the jungles. The immediate enemy, though, was fierce mosquitos. Netting was finally obtained and set up around the bunks.

BOMBER CRASH

Tuesday, August 25. Tragedy struck on an island outpost with the crashing of a B-17 four-motored Boeing Bomber last night during a fierce, tropical monsoon. The fatal accident occurred at approximately 2130 as the plane was returning to home base after a bombing mission about 600 miles off-shore. Five crew members lost their lives in the fatal crash as the heavy bomber was unable to climb out of a landing glide. It crashed into a rocky hillside about 50 yards off the apron. Three of the crew members were thrown clear and a fourth was carried from the burning hull just in time.

"Radio Silence," an army order, was blamed for the fatal mishap since the pilot could not determine the lay of the field through the heavy downpour and with a zero ceiling. A poorly placed bank of flood lights was partly responsible factor, with the light rays shooting diagonally across the runway.

MONSOON

For fifteen minutes I quietly listened to a sprinkle of rain and delighted in its clatter. Suddenly, a boom caused by a tropical monsoon shattered the air and a flooding fury of rain poured and poured for about 45 minutes. I was fairly dry beneath my rain poncho but soon made a mad dash for the tent only to find a puddle of water at least three inches deep under my bunk.

Later that night, and after returning from the air strip at midnight, I remade my hammock bed (using a slightly damp mattress) and slept until morning. When I awoke, I found my seabag, with contents soaked, lying in a pool of water on my cot. My M-1 rifle had six inches of water in the barrel. A portion of the next day was spent in drying out and caring for clothing and gear to prevent it from becoming unusable. The morning report from the airstrip: Two Japanese carriers on fire and one Japanese carrier sunk during the night. All bombers returned to base—only to lose one on the runway crash.

350 PIES

Whenever the rain continued for more than a few hours it was almost impossible to dry out. Nearly everything rested on the damp ground and periodic head colds and bad coughs were common. Today I helped Pop bake 350 individual mince pies and was able to make myself a little mince jelly and tarts to go with it. During the day I sold a fishing reel and some line to Jewel, a black steward, for seven dollars in cash and the promise of some food favors at a later date.

In spite of working very long hours, whether feeling sick or not, there were some occasions for leisure. On Sunday I again helped Pop bake chocolate pies, put together a few canned strawberry tarts, and we had a great dinner of potatoes and fresh liver. Canned goods, especially cans of peaches, were plentiful and contributed to occasional feelings "of the good life" on a remote, tropical island.

The heat and the humidity continued to be very intense. Chief Ray discovered a four-foot long snake in camp last night and put a sudden end to it. Next day, several of us started brewing some beer and tended it over the next three days. Some of the 6th Battalion moved into our camp. More were expected to follow in about ten days. Tensions were high. The weather was very hot and I started wearing a .22 pistol that day.

MAHOGANY, ROSEWOOD & TEAK

Monday, August 31. Much of the day was spent in cutting native woods. The most attractive hardwoods flourishing in this tropical environment are mahogany, teak, and rosewood. The mahogany was used for fence posts. Teak was exported for unknown purposes, and the rosewood trees seem to grow just as fast as we cut them down. Rosewoods have very unusual growth patterns. Many times their branches are larger than the trunk of the tree. Rosewood grain patterns resemble bowls of spaghetti—with color variations that range from definite red to many shades of brown. Teak wood, in contrast, is light brown and has close grain patterns that are full of almost microscopic specks of oil. Teak is an extremely fine wood for hand carving.

As a diversion from the regular routine, a little time was freed to cut stencils in thin pieces of teak wood. Tee shirts could then be stenciled with the names of our previous bases. Our own unit, for example, started at Great Lakes, then Norfolk, Point Hueneme, Pago Pago, and now, New Hebrides.

WILD PIG

On the following day, I visited the 98th Bombing Group's Army Air Force galley. There, some of the men had brought in a wild pig for the cooks to barbecue. The freshly slaughtered pig was quite a treat and was fixed up royally by the appreciative cooks. One of the cooks also was quite talented in making baking powder biscuits. Along with fresh slices of ham, the hot biscuits and jelly were a real delight for the airmen. There was a large number of airmen at the base after being forced to fill the airfield with planes off the sunken carrier, *USS Hornet*. The airmen were assisted in setting up tents and in getting them regular meals. It was one of those rare opportunities for land-based Seabees to socialize with combat flyers and to hear first-hand of the great sea battle just ended.

TORPEDO DIVE BOMBER

September 6. On this date I had the opportunity to ride in the Torpedo Dive Bomber. The visual experience reminded me of the vastness of the Pacific Ocean and the extreme smallness of our island group, located thousands of miles from home.

Each day's work started before dawn and ended after the clear sky's sudden tropical darkness had settled with complete silence around every object large and small. Exhausted, sweaty and dirty men, unable to see well enough to continue working, piled onto unlighted trucks and inched their way either up or down the airfield to carry them back to a dimly lighted mess hall.

"What's for chow?"

Same old response, "Australian mutton."

"Hey, Pop Nelson made some pies last night." And once again it was the little surprise that elated the weary men and offset the consistently low quality of the main meals.

DAILY LIFE IN PARADISE

Day after day, the camp was enlarged to accommodate more men, equipment, and supplies. At the same time, others were being transferred to a new camp where their particular knowledge of construction could be used.

Our equipment on this immediate job included an almost worn-out, medium-sized road roller, a road patrol machine, and two dump trucks. A nearby army unit gave us the grader equipment and we begged for an additional dump truck from another army unit. The heavy equipment was skimpy, but with a daylight-to-complete-darkness work schedule our progress was only a little short of miraculous.

Our camp boasted a grass hut, or "native fola," an oil-heated hot water still, and improvised showers for the men. Since ours was the only hot water shower for miles around, there was always a waiting line. Men from the nearby army bombing groups and the marines from the coast patrol enjoyed their first hot water baths in several weeks. The still also provided pure water for drinking and cooking and the much needed distilled water for the blood plasma transfusion unit at the nearby field hospital.

With the coming of the monsoon rains work slowed up a little. The Seabees, busy assisting the marines in stopping Tojo's march south, kept up a full work schedule rain, or shine. Shortly after each tropical rain, the heat soon dried the men and allowed more work to be completed. At times, shelter from a particular downpour might be a plane wing, a crushed truck cab, or a clump of trees. Whatever the exterior protection, clothes were often quite damp or soaked regardless of the amount of rain because working bodies produced enough sweat under a tropical sun to be soaking wet anyhow.

As the airfield lay at the base of a small hill, the

monsoons quickly produced a small lake. Suddenly, out of the clouds, a heavy bomber would appear and begin circling for a landing, level off, and ride the runway into the newly formed lake. Water sprayed for two hundred yards on either side. Again, workmen were prone to be drenched. However, almost as quickly as it formed, the new lake on the airfield would disappear into the coral sand and the runway was suddenly dry again.

Available machine operators and Seabees would run to the landed plane to lend a hand in assisting any wounded crew members. At times red blood could be seen dripping from a number of anti-aircraft bullet holes or flak damage. Uninjured crew members were often quick to examine the flak holes and other damage to the wings and hull of the craft.

Soon, a second plane would appear on the horizon, wheels down and coming in low—an obvious indication that their gas was extremely low. As the plane neared the field someone shouted: "She's coming in on two engines. Watch out for a crash landing!" Somehow, unbelievably, the pilot and co-pilot would set this bird down without a mishap. Again, the available men would run to the landed plane to assist and ask questions.

"The battle is a bitter one," replied the co-pilot, "What about my men?"

The injured men were already being helped. The ambulance was already handing out a stretcher, and the blood plasma was being connected to save another "fly-boy."

ON THE GROUND

"Well, men, that makes ten of them back. I wonder where *Yankee Doodle* is at this time. They should have been back by now."

"Right," said one of the nearby flyers, "there were eleven in the flight this morning." With this question going from parking place to parking place, the exhausted airmen soon faded from view as they walked quietly along a jungle trail on the way to supper. The eleventh plane didn't make it.

Supper that night was sandwiches, cold tea and dessert from a large container of "K" rations—a chocolate bar, or dried up prune bar. At the airstrip, the "mecs" began servicing the planes for the following morning's dawn patrol. There, the "mecs" who wanted to could treat themselves to a small refreshment—a very small "Pink Lady"—to cut the dust in the throat.

Again, all was quiet, and the Seabees returned to building a safer field for the Army Bomber Command. The extreme darkness of the tropical night was very effective in concealing each night's construction activities. Each new morning, the day shift was impressed by the amount of work accomplished by the night crews. Also each night our final day's news was summed up by the crew members as they gathered at the showers, or went through the Seabee chow line.

The *Yankee Doodle* had come in late. The crew had been watching a Japanese convoy that was steaming south past the Solomon Islands. "Looks like a little action for us in a few days," said one of the *Doodle's* crew. "Should take them three days to reach this island, so we have plenty of time to prepare before the fireworks start."

Again, the camp scrambled in preparation for the anticipated action. Foxholes were hastily dug and rifles cleaned and oiled. In many cases, the "foxholes" had been used for a different purpose. An earlier order had been given that required the digging of slit trenches. "Traditionally," the Seabees were not much on taking orders that involved work with little immediate purpose of some sort, so the supposedly complete trenches were typically near shallow spaces on the sandy jungle floor and heaped up with leaves. Now, it was different. A Japanese battle force was approaching this island directly. Tents were moved back another fifty yards inshore. Dig! Dig! Dig! Quickly the shallow recesses became sizeable enough for at least four men each. The fast work was a tangible example of Seabee "Can Do," as everyone pitched in, and each protective space took on a slightly different appearance.

Most of the rifle pits were covered with steel matting from the air strip runway. Mine was mat covered, burlap lined, and stocked with food and water. This was a tee-shaped trench, readily covered by merely jumping out of the tent into a slit trench about four feet deep. The objective then was to crawl under the protective steel matting and back into a shallow cave lined with burlap and covered with at least a foot of sand.

Men gathered in small groups after completing their preparations to discuss high hopes of their chances to "get a couple of Japs" and do some psychologically valuable souveniring. Jokingly, and for special effect, someone shouted, "Don't forget, Jap ears are worth ten bucks a piece." Someone else added, "Sure, remember how some of the prisoners in the stockade covered their ears with their hands when we mentioned this to them. One of the prisoners could speak excellent English and had been a graduate from my old school—the University of Oregon. "Then they would hide under the brush bunks," said another guy with a very serious tone, "and I bet those ears won't last long once the marines return from the Canal."

BACK TO WORK, ALWAYS BACK TO WORK

Work details went away to their various assignments, but none were as happy as the men on the bomber field. The reason was simple enough in a region bereft of information about the outside world. The bomber field crew was always the first to return to camp with the hottest news of the day. Our job was to finish the air field as quickly as possible. Our crews never stopped working—morning, noon, night, rain and shine.

With every moment of daylight, the grader stopped only long enough for a circling plane to land or a violent monsoon to blow over. These monsoons would stop the engines unless the electrical system was waterproof. During the early stages they weren't.

Endless work was done beneath the hot tropical sun and wishes turned to cold beer—but there was no beer to be had. Mostly, work consisted of removing the top muddy soil and then refilling with coral rock and sand. Rough grading was done by hand, followed by a road patrol and light grader. This surface was then covered with the large metal strips which create a durable runway regardless of of the weather. The metal mats were two feet by eight feet and considered a most valuable piece of equipment. Fine coral sand was then spread over the metal grates to prevent excessive slipping. With this process, the Seabees completed reliable runways in a short time and mostly used available materials. This was the labor that was beating Tojo to the punch on his own land. Less than twelve months before Japan had been practicing war games in the New Hebrides. Their previous occupation of these islands made us feel the same islands would again be targets in the expected southward advance of the enemy.

By the noon hour of this day, talk had drifted from the on-coming task force of enemy ships to remembrances of stateside experiences.

"Sure would be nice if we could get some mail around here," said one of the guys. "Seems funny that no one is getting any letters."

"Yeah," said another, "Six weeks out and not a single line."

One of the youngest guys in the group said, "Yeah, hope my mother's doing okay—she was real sick when we left last month."

"Hey, I'd sure like some candy or fresh fruit from anyone!"

"Sure," said one of the older guys, "No candy, no fruit, no beer, no cokes, no women and NO MAIL."

"Hey, look out," shouted another, "that cockeyed marine fighter pilot over there. What's he think he's doing over this field?"

"Guess things are quiet enough over the Solomons that he's got time to scare us to death."

"Hell's fire! That guy's less than ten feet above our grader. That jerk could tear up some of our equipment and people along with it."

"Get down, hit the deck!"

"He's doing the exact same pass, he can't always be so lucky with our lives."

[These pilots were stationed about three miles north of us on the same island. After moving to Phoenix, Arizona, many years later, we became friends with a marine fighter pilot who had been stationed on that very field.]

A detachment of the battalion, Company A, was located on the Frenchman's plantation and was working on a dock as well as building an air strip for the Marine Fighter Squadron. A friend, Leo Hatch, a cowboy from Montana, was in this company. He was one of the few men who had ammunition for his rifle. One day he came by my job and explained how mad he was at the Frenchman
(the Frenchman had not been moved out yet) for his treatment to one of his slaves. Leo had armed his rifle and waited several hours to shoot the tyrant. Luckily, the Frenchman did not show up or there certainly would have been lots of deep trouble.

It was always hard to become accustomed to the tropical days. The late September sun set about 1800 hours and it would be pitch black within just a few minutes. Fortunately (for our sanity), there were occasional films to relieve the monotony of work, eat, and sleep. The titles and subject matter were unimportant. Just something from home. Another plane crashed yesterday.

Was especially glad to leave this island behind. At least it was still the winter season; otherwise the heat would be unbearable. The harshest thing to deal with at that moment was the non-stop wind. Spent the day rough grading the airstrip and at the same time fighting the heavy winds and brutal dust clouds.

NOTHING NEW NEVER IS.

Mail call. Received five letters from home today. One of the letters was from the factory worker in West Virginia. She was the one who waved at our troop train as we traveled from the Great Lakes to Norfolk. She shouted her address down from a fourth story window while we waited for a train to pass. Our correspondence continued for some time, but we were not to see each other again. The mail was about two months old but no cause for complaint. Sharing parts from each other's letters lasted for some time, and we stretched the pleasurable time to the max.

NEW ZEALANDERS

A detachment of New Zealanders moved in on the bomber field today—the Royal New Zealand Air Force. Here was a group of men who had seen much service. They came to the South Pacific from the Mediterranean Sea and had been in action in North Africa. This was an opportunity to spend some friendly time swapping stories with the new men and swapping various articles of value. They set up their own galley, and there were good opportunities to talk with the cooks and share meals at any time of the night or day. The New Zealanders were a curious lot. Many were young boys in their teens; the "old" men showed much evidence of wear and tear. One man had a deformed leg and a distinct limp. Most looked as if they were just about to pass out.

The Royal New Zealand Air Force had been sent in earlier to add support to the sea patrol. They were based a couple of hundred yards up in the jungle from the head of Bomber One. These men were very friendly and were able to tell of experiences in North Africa and other places along the way. Their tents were pitched in the mud and rugged living conditions were the same as the 7th Battalion. When possible, we would trade items with them and they would use our showers. Their commander was giving Admiral Bill "Bull"Halsey the material they had shipped to Espirito to build living quarters. We began using the material to build a designer house for the brass.

As I tallied the truck loads of lumber out of their yard, different men would ask where and why.

"This stuff is going to build the admiral's quarters."

"Hell, the bloody left-tenant promised us it was for tent walls and walks."

So I said, "You want a board, take one."

Each new day the fighting men from New Zealand asked the same question and I replied, "Take a board or two." Soon the men were able to walk between tents and have dry feet. The satisfaction of being able to assist fighting men in keeping their feet dry was immense. The admiral's base eventually consisted of two regular houses.

SAME AS IT WAS

This day started much the same as the others had for the past several months. A dull breakfast, maybe a new ship in the lagoon, and the same pounding sunlight. Sid had won a cool hundred playing poker at the hospital the night before. We had to hear all about it as we loaded into a troop-carrying truck on our three-mile ride to the dock and warehouse area. Hey, it passes the time.

I was feeling fine and figured I should do a little extra labor for the war effort. My job was to strip some footings, one warehouse to be erected, and I was swinging away with the sledge hammer, knocking the wood forms away when, whammo, damn! Something hit my face. My mouth started bleeding. I felt bits of tooth; I sat down and spit more blood and pieces of tooth. After a few minutes and after washing my face and mouth, I learned I had knocked out one of my own front teeth. The insult to my mouth was bad enough, but the extraction of the root by the camp dentist, without Novocain, was real pain.

For no obvious reason, my mind drifted back to one of the first days after boot camp. The lieutenant said, "Go over there and build a loading dock for the garbage cans."The lieutenant proceeded to tell another platoon to start work on a baseball diamond. For a few moments, my platoon studied the loading dock idea and decided it made more sense to work on the ball field. We did. The lieutenant later reappeared and inquired about his original orders. "We thought the ball park was more important." He agreed and the garbage can dock was forgotten. Life was simpler then.

Weather today is overcast and the flies are biting. Must be about ready to rain. The report in today from the crews of the TBF indicate the Japanese are taking a breather at Guadalcanal. Additional reports are coming in from Schenectady, New York, via station WGEO, the General Electric station. Also, some reports from Australia. Same thing. The immediate concern was with the water supply. The drinking water is especially bad. No water treatment equipment in yet. Almost certainly, it will rain again tonight.

Another tropical monsoon last night. Lasted several hours. With time to reflect on the "state of the world," it seems that thirty years from now every detail of this present life will have more meaning for a single citizen who loves liberty and is willing to sacrifice for it. Back to the present, the local 98th Bombing Group lost three planes in the past week during a fight at Bouganville, the Japanese held base in New Guinea.

Relocated the galley this evening. "Pop" Nelson, a wily character from Portland [OR], was an extraordinary head cook. In reflection, the galley was indeed a fine place. The soldiers, marines, and displaced navy fliers kept the chow lines so long that we had to stop the local natives from eating out of the galley at critical meal times. Typically, the black native contract workers would pick large green leaves for plates, then go sit together in the shade of a palm tree to eat. The chow wasn't very appealing, but you wouldn't notice it based on the native's appetites. The galley was reserved for military personnel only.

DAILY GRIND

The army mess hall furnished quinine tablets to every man. There were always large bottles of the tablets sitting on the tables as a reminder. The first time I experienced quinine was quite by accident. The weather had been quite hot and I sat down for lunch across from the Bomber One strip. My attention was on the work we had been doing. I popped a couple of quinine tablets in my mouth thinking they were salt tablets. Wow! Big mistake. The mess sergeant saw what had happened and insisted that I take the tablets on a regular basis; consequently, I think that was the reason I did not catch malaria.

On another visit to the army kitchen, we had the opportunity to barbecue a wild pig complete with dumplings. The Seabee cooks gave the army cooks supplies they could not obtain through their own sources. I had the good fortune of working in the galley at that particular time and was, therefore, one of the first to eat the most.

Most of the time there was little to be truly excited about. There are few instances of what might be called high drama, except for the unavoidable plane crash or stories of action from the various fronts. Life has a repetitive pace about it that is similar to the waves breaking over the coral reefs. The little things have a way of taking on great importance. Canteen supplies have been running very low. No candy or gum for several weeks now. There are few books around that haven't already been read, or films that haven't been seen, repeatedly.

We did have hot water for showers—if we were lucky. Next to good food and an occasional beer, the fresh water showers were intensely popular with all of the various units stationed for miles around. The socializing in the shower line and the shower itself was something to be relished after intense work in the tropical heat. The shower line was therefore an ever-present sight 24 hours a day.

It had been almost ninety days out of the the United States. Often, a very disagreeable three months. I had taken almost that long to acquire the materials and assemble a small writing desk. Finally, after more than eight weeks of living on sand and dirt floors, we were issued enough wood to build tent decks. The men in our tent put together the decking in a short time and at the same time put the sides of the tent up for better ventilation and light.

The next day orders came to dig foxholes. The response was barely noticeable. On the day following, a direct command to dig foxholes came in and the men fell to with shovels in hand. The orders seemed absurd considering the fatigue of the men and the absence of any air alerts. However, within a couple of nights and just after the sudden darkness, boom! boom! whish! whish!, bang! A Japanese submarine offshore was shelling us. Fortunately for us, the sub was traveling and it was dark. For obvious reasons they could stay surfaced for only a very short time and did not inflict any real damage. But on the following day the dirt began to fly early. Small, shallow holes dug earlier were now dug down to be two-man foxholes complete with metal runway mats used as roofs and covered with dirt. My own foxhole was a small Hilton—a slit trench at the edge of the tent with a tunnel off to one side, metal mats for the top and lined with burlap and stored with food and water. I actually slept in there on a few occasions because it was cooler.

Pay day today, but not much to spend it on. Save a bit, gamble a bit and send a bit home. The big news of the day is a report going around that the first bombing of Tokyo has been pulled off with 18 B-25's from the carrier *USS Hornet* from a distance of 600 miles offshore [Gen. Jimmy Doolittle's Raid]. The bombing was a great morale boost for all of the guys here.

On the same day many of us were able to inspect a Japanese camouflage suit, a souvenir of M/S Leep, Comstock, Nebraska. He also had a .25 caliber Japanese rifle that had been acquired following a battle on one of the islands. The threat that a group of Japanese Marines might come ashore and launch a late night assault was an ever present thought. It was an especially powerful thought when we occasionally had a moment to swim in the surf. Of course, the real threat was the possibility that a white shark might appear. For that reason the local natives rarely went into the surf themselves.

October 13, 1942. Have been having trouble with the lights for several nights. The electric crew seems to have gotten the portable generators working again. We're back to having one light bulb per tent and, with effort, can read, write letters, or at least see each other's faces while talking. Used the opportunity to see to blank out any military statistics that might have accidentally been put in my day-to-day journal. Certainly a shame to delete, omit and conceal the most interesting aspects of life that filtered into the isolated outbacks of the war. The appetite for the latest "news" of the war was always intense and could have life and death consequences, such as digging foxholes *before* we're shelled by a Japanese submarine. How many times have we heard the expression, "Loose Lips Sink Ships"? Even memos could have been more specific. The written guidelines for what was forbidden or not forbidden to write down were often vague. The general practice was to typically write down whatever you want and then edit out anything that might in any way be useful if captured by the enemy.

On the next day, I had the opportunity to talk with a soldier from the state of Georgia. He delighted in telling of a company there that was experimenting with a glass tube two miles long that was used for transporting chemicals. In the process, the chemicals were "cooled" with live steam (boiling water and air temperature water to prevent the gigantic tube from breaking). Somewhere the logic and purpose behind the glass tube was lost, but any details, other than island details, could be quite fascinating.

One month later we were stationed along the fringe of the jungle and close to the beach. Work had been steadily progressing on "Bomber One." Few of the fellows seem concerned about possible danger from the Japanese. This strange sense of security was in lieu of the fact that the Japanese sank the carrier *Hornet* at sea and forced her planes and men to join up with the B Company men. The intensity of making preparations for these additional men added substantially to the daily work loads.

TRICKS OF THE TRADE

A makeshift Thanksgiving was celebrated along with the men of the Navy Advance Base. There, working in the CPO mess, we learned a few more tricks and angles needed for survival in the military life. For one thing, it was discovered that a wide range of fresh fruit and other items could be procured after getting permission to visit a newly arrived supply ship. Initially, I went aboard to purchase apples and oranges for the company mess. While aboard we were offered a sit-down meal of ham, eggs, apple pie, and a bit of stateside news. We bought a total of four cases of apples and oranges for thirty two dollars and made a lot of men happy in the mess hall to get something fresh that hadn't been grown on the island.

The Christmas and New Years seasons passed with little notice out of the ordinary except for some occasional cards that were received during that time. It was almost the end of January before the next batch of mail was received after the "holidays." Had breakfast aboard the Seaplane *Tender Curtis*. It was memorable because there was typically no shortage of good food aboard the ships. One particular supply ship always had a couple of dozen fried eggs, plenty of toast, and hot coffee every morning. I had the good fortune to be selected as a purchasing agent for the ship's store, so I had passes aboard every ship in the channel at one time or another.

Most of the supply ship personnel were really friendly, others were more exacting. One experience I can recall involved the dress of the day. The *Curtis* was usually quite friendly as they were anchored in the channel for extended periods while maintaining a land-based repair shop to service the flying Catalinas. Our outfit laid the metal runway mats on the sandy runways that allowed the Catalinas to come ashore. The *Curtis* had an open work-boat to transport from ship-to-shore and vice versa at regular times each day. It was fairly easy to learn the schedule and make arrangements such as buying ice cream from the ship's store, buying a little "medicinal" alcohol, or working in a haircut, etc. Occasionally, it was possible to buy or barter for pies, fresh fruit, and once, even for plywood from the ship's carpentry shop.

On one occasion I was stranded on a tanker for at least seven hours, had lunch with the crew, and finally got a ride to the beach on a liberty boat. Casual arrangements were made to take the "off-duty" crew members to our theater tent, known as "Big E," to see a film. The pilot dropped us off, and we had a very wet walk back to camp to see a film with the enthusiasm of school boys.

At another time I was aboard the *Minneapolis* while efforts were being made to repair up to sixty-five feet of bow that had been shot away. Various Seabees and the ship's repair crew were welding a temporary break water bow in place of the damage and preparing for a run to the shipyards at Pearl Harbor. It was necessary to return for more steel before sailing forth and arriving safely. While aboard, I bought some officer's insignias. I later sold some eagles to the new captain of the island. Commander Piland called me on the carpet for that little transaction.

During this slow period of "goldbricking," I had built a footlocker for an army lieutenant. The material and shop usage were provided compliments of the ship's carpenter shop aboard the *USS Tangier*, another seaplane tender with several close friends on board. Earlier, while living at Bomber One, another fellow and I had built up a custom tent for an army lieutenant in the camera department. The effort helped to establish good connections with the small photo lab and was a source for photos of points of interest around the island and of ourselves.

OUT AND ABOUT

Many of these lower priority activities were initiated through the Port Cargo officer. There I met a lieutenant commander from Seattle, Washington. He had been a personnel officer aboard a ship before being stationed at this small island. Coming over, the Japanese air raids and other attacks had scared him intensely. He virtually could not go below the ship's top deck. The solution was to transfer him to a land office. He was quite a sensitive man and showed respect and concern for the men about him. His office supplied me with water transportation around the bay as needed. Periodically, he would also loan me his jeep saying, "Take these papers, and if stopped by the M.P.'s, tell them you're a messenger. Otherwise, just discard them." One day he asked me if I could build a footlocker for a Mr. Underhill, civilian advisor for inter-island shipping attached to his office.

"Yes, sir, if I can find the plywood. I'll also need your jeep for transportation." Several days passed before I noticed some plywood next to the battery shack. I checked around and then asked the headquarters man who owned it.

"The chief stole it off a load," he said.

So I said, "I'll be back this afternoon to borrow a couple of sheets myself."

I rushed to the Port Cargo Office and obtained permission to borrow the jeep. It was then a simple matter to return and quietly pick up two sheets of plywood and transport them to the "Net and Boom Detail" of the ship's carpenter shop. Building a footlocker was easy enough to do after that. It was just another small and needed favor that made it possible to chase about in the Jeep and stay active during the slack times.

The headquarters man at the battery shop came through the chow line, and I gave him two extra, cold cokes to go with his lunch in exchange for the plywood. Later, Mr. Underhill, the civilian, provided a couple of highballs and a quart of whisky. These were sold to a First Marine Raider for roughly forty bucks, and the money was given to the headquarters man for payment in full.

The weather continued to be dry and windy for the past ten days to two weeks. The time either drags or flies depending on your state of mind.. We were bombed by the Japanese twice last week with no damage, as usual. Sometimes it seems as if they're just going through the motions to satisfy a remote commander and not be readily killed themselves—by either side.

CARROT AND STICK

Lieutenant Tinsley circulated a little scuttlebutt that we'd be in New Zealand in four to six weeks. No truth in that one, however. Instead, it was continuous downpour of water from the local monsoons day after day. Too wet to work on anything except the most important projects. Made the time go much slower. Often, the men simply turned in early because it was too wet to do anything else. Rain, rain and more rain. Time to read but a real scarcity of new mail. My personal cash on hand is slowing building up to almost $300, but what's money with no place to spend it?

Wednesday, January 27th. Worked for Lt. Tinsley throughout the afternoon. Afterwards, he recommended me to a lieutenant and one of the base captains with special projects in mind. The carrot and stick offer was a suggestion of transfer to the boat pool. My response at the time was the traditional, "Sure hope it goes through." Such idle dreams in a war zone can either build up morale or cynicism, depending on circumstances. Such a transfer would have been a real change from the Seabee daily diet of cement, dirt, and sweat.

A few weeks and no transfer later; a request was received to report for work to Chief Bolt. There had been a time when this same man would not allow me to work for him. Now, for whatever reason, he had an urgent need for carpenters. So off to work I went. One of the thoughts that occured to me at the time was, "If my recommended commission doesn't come through because of this guy, I'll be damned if I'll work for him."

It was just a couple months back that a similar incident occured. That is, I had done extensive work for Lieutenant Canting, off the seaplane tender, *USS Curtis*. He had been attached to the photo lab aboard ship. He wanted me to obtain reassignment orders and fly to Guadalcanal and build special tents for him and his photo recon crew. But, without explanation, Lieutenant Tinsley would not let me go. Naturally, my estimation of him sank lower than the surrounding coral reefs. So it goes.

Thursday, January 28th. Began a whole new project that helped to take my mind off the carrot and stick promotion. Today we were sent to start construction on Admiral Halsey's headquarters. Here, for the first time in several months, I was able to work with my old gang [Actually, I later saw several of these men during a 1972 Seabee's reunion in San Francisco].

BUDDY SYSTEM

Sunday was another unusual day during a typical day of November rains and round-the-clock runway construction. A driver to haul crushed coral was needed. Being recently qualified, I was issued a beat-up Ford dump truck of ancient vintage. The job was to obtain loads at the crusher and then deliver and spread the coral on the wet spots of the runway. Seemed simple enough. However, after several trips, the chief stopped me and inquired, "Fay, you know where Chief So-and-So is located over in the jungle?"

"Sure."

"Take the next load over to him and dump it wherever he wants it."

Upon arriving at the instructed spot, I found a chief in charge of a navy mess hall, or maybe it was a a marine mess hall. Anyway, I dumped the coral in the walk next to the entrance of the mess hall. The entrance had been covered with several inches of water and large standing puddles. The chief said, "It's the Old Navy Way. You do for me and I'll do for you."

His attitude was friendly enough to encourage me so I said, "Chief, you need another load."

"Can you bring me one?"

"Only supposed to give you one, but for two cans of number two and a half peaches, I think I could manage."

"Okay, when will you be back?"

"Immediately." And just as quickly I had two cans of delicious peaches for my effort.

"Chief, you need another?"

"What? And for more peaches, I suppose."

Once again I had more peaches. The improvement for the enlisted men's mess hall was substantial. For me personally, the effort was far more satisfying than adding two unnecessary loads of coral to the runway. A good motto for the time was "Little favors that will benefit many."

That November was an extremely busy month and never slowed down. The food improved along with our increased desire to eat whatever they served. The PX started selling candy and the accessible areas of the island were becoming more familiar. Quick dips in the coral sea were becoming more pleasant because we could now rinse off the salt under our fresh water showers. The Japanese were apparently being detained somewhere to the north and attack incidents were becoming quite rare. Mostly, it seems, we were all becoming accustomed to living under the "worst of conditions" with some dignity and small comforts.

TORPEDO BOMBER

One particular highlight for me was a ride around the island in a navy torpedo fighter bomber that was conducting regular shore patrols. A chief, who was friendly with Pop Nelson, offered to take me on board. He gave me a steel helmet in place of an aviator's cap. I was technically signed on as a gunner but had not been instructed in the use of the 50mm cannon on board. The steel helmet was issued for two purposes. The first was to wear in case we were fired upon, and the other—you guessed it—was to hold inverted between the legs just in case. Fortunately, neither usage was activated on that day. This particular plane had fold-back wings for compact storage below decks. As we were taxiing down the runway and simultaneously trying to lock the wings into place, there were three other similar planes in our formation who were already sky-born. Finally, after several tries, the lights indicated that everything was a go. We were seriously on our way into the "wild blue yonder" and quickly rendezvoused with the others. The noise and acceleration were quite impressive. The flight around Espirito Santos was one of the most interesting two hours of my life. We spotted a shadow in the blue water and upon closer investigation discovered only a mat of seaweed, not a lurking enemy submarine. That particular island was so rugged that, had the Japanese landed ten miles away, it would have taken them three weeks to penetrate the jungle and steep mountain sides. Some of the mountainous cliffs were sheer rock formations that were at least 1,000 to 1,500 feet straight up from the pounding white surf below. Oddly enough, there were actually high plateaus that were suitable for grazing small herds of cattle. We circled over one small herd and they stampeded wildly to get away. The deep, mountainous canyons had a streak that appeared to be a silver thread winding among vine-covered walls and had the image of an exotic fairy tale. For the benefit of his passenger and bystanders on the ground, I'm sure, the pilot climbed to 12,000 feet and dived down at a warship anchored in the harbor. The ominous sensation was less than pleasant. The plane appeared to be going straight down; at times the body of the plane shuddered, and so did I.

GOLDBRICKING AGAIN?

Some days the ambition to work just isn't there. We were all tired from the long days, repeats of the same lousy food, lack of nails, tools missing or broken, poor sleep in the high heat and high humidity, and contradictory orders that make little sense out of whatever has to be done at the moment. Sometimes concern about quality of the work done fades and even disappears for a while. Some of the fellows claim they haven't had a day off in several weeks. Besides, what was there to do on a "day off?"

February 3rd. No rain tonight. I spent the last few days working for Lieutenant Tinsley. Just another goldbricking job in the absence of real work. Plenty of reasons to continue requesting a transfer to the admiral's staff. Obviously, the request for transfer was not going anywhere because "they needed carpenters."

Enough repetitive, grueling work and just about anyone's interest in commitment to doing a job can wane. The isolation and deprivation of nearly everything except the basics to survive simply becomes boring and irritating. Most of the men eventually gravitated toward a particular diversion. Some played cards at every opportunity, others drank beer and tried to outdo each other in telling lies, some worried about their wives, children and parents back in the states. Concern about receiving a "Dear John" letter was a reality for many men. Almost any bad news from home would just have to be coped with at a very long distance. Some guys passed away idle hours playing ping pong, other ball games, or just reading scarce books and magazines in the rec hall or their tents. My particular diversion was to concentrate on new angles for putting trades together—the old idea of small things that do a large amount of good. Today, for example, I spent personal time laying linoleum on the tent deck at the Port Cargo Office for Lieutenant Tinsley. The access resulted in my having the opportunity to dine aboard a sea-going, AV8 vessel that included cold jello, hot biscuits with honey and all the trimmings for a gourmet meal.

ISLAND PALMS

Extensive time was spent working on the headquarters camp for Admiral Halsey. In retrospect, and considering the events that followed the construction of several headquarter buildings, those days were among the most pleasant days of service while stationed on that pile of coral rock. One of the first tasks was to cut down a number of coconut trees to create a clear view of the harbor from the admiral's porch. Each tree felled cost the U.S. government a certain number of dollars that had to be paid to the island's natives.

As bad luck would have it, the first tree cut fell across the hood of Seaman Mann's dump truck and created a festive mood among the fellows. I can't remember if the truck was still operable or not., but there was substantial damage to the water pump and radiator. Each tree provided a delicious, candy-like substance that was very much like divinity candy. To this day I can't say what specie of tree it was.

We finally started work on the foundation and were obliged to mix the cement by hand with shovels and hoes. The reason for doing it that way was simple enough. The lieutenant in charge didn't make the small amount of effort required to secure a mixer that had been sitting idle on another part of the island. It was often the "little" things such as the mixer incident that bugged the men doing the bulk of the physical work and chipped away at their morale. The officers in charge could be quite jaundiced and missed opportunities for providing moral support by their failure to see to it that there was fresh fruit in the chow line, or enough butter served, or even enough food to adequately feed working men. Ironically, adequate supplies of food were available from the depot—fresh apples, cabbage, potatoes, and plenty of canned and fresh fruit. Yet the word going around was that the commissary wouldn't spend the money to keep supplies adequate. One small comparison: Men in the large army mess hall had jellies by the jar, and we did well to have jelly by the spoonful, if any.

Wednesday February 10: Started off the day in a delightful mood after a good night's sleep and actually had vivid dream of being back in Oregon with an old friend. Also, received some mail today for the first time in awhile. The mood was dampened soon enough by the first rain in several in days. It was almost time to acknowledge George Washington's Birthday and various men were receiving their first away-from-home Christmas cards. Cheerful cards were a visible boost to despondent Seabees. The risk of receiving a "Dear John" or belatedly learning of a death in the family built up great pressure because of the long intervals between getting mail and the extreme distances involved.

LUMBER TALLY

Valentine's Day, 1943. Several of us were assigned to the lumber yard at the Royal New Zealander's Air Force Camp. The purpose was to tally lumber that would be used in the construction of the admiral's headquarters camp. Procuring items as simple as 2 x 4 studs was a major struggle. The troops there were cooperative but a beleaguered lot. They appeared either older or younger than most of our forces. The apparent reason was that New Zealand had been in the war for years when we arrived, and their manpower was being depleted rapidly. At this particular camp, however, the situation looked pretty good. There was a large and well-stocked mess hall, a recreation building, a fancy officers' club and tents set up on wood decks for maybe 200 men.

Several weeks elapsed and I barely noticed the days going by, only the new date. On Sunday I was supposed to have the day off, but was sent on a rush job. It's always rush to do this and that, yet we still can't get the 2 x 4 studs that are needed for the admiral's headquarters building. Men are still working on the coconut trees so the admiral and his staff can more conveniently see the harbor.

DRY CANDY

More days and more rain elapses. On this day I went on a solitary hike along inland trails for the better part of a day. It matters little that it rained periodically. By now it was just accepted that rain means almost nothing. We get wet, the heat dries us out, it starts all over again. The loosely planned hike started out with a problem. The water was up on a local river and there was no bridge or boat available. I had filled my ditty bag full of candy to encourage some trades with the natives. There was only one possibility left for me, and that was to swim. It didn't matter that I got soaked to the skin, as long as the candy stayed dry. Finally, after much walking, hacking through jungle undergrowth, swatting at bugs and wiping away sweat, I found a small native camp. There I was welcomed and offered a late lunch with the native black fellows. At almost the same time, a monsoon let loose and it seemed there was up to five inches of rain in less than one hour. Again, soaked and trying to keep the candy dry, I learned the native camp I was looking for was farther around the shore line, so I decided to return to camp without making any of the hoped for trades. So it goes.

To get back to camp, I now had to swim and wade through two swollen rivers. I was completely soaked for the fourth time today and still had one more river to cross. The temporary wetness was not so unpleasant as it might seem. The natural beauty of the island itself is all pervasive. I had lots of private time to reflect on my current, youthful adventure, and time to eat seven candy bars that were no longer part of the "trade goods."

March 6th, Another week has passed almost unnoticed. Work continues on the admiral's quarters; including two-and-one-half-inch water lines in many directions so the senior officers are not inconvenienced by standing water after each monsoon. The framing was going along well. I looked out one of the recently framed windows and spotted a familiar face among several new recruits walking by. One had the name Keebler written on his dungarees with bleach, the navy way of identifying clothing. I yelled over to him, "Hey, Keebler, you from Oregon?"

"Sure thing, how'd you guess?"

"You from Lebanon [Oregon]?"

"That's right."

"You have a brother, Don?"

"Sure have."

The excitement of seeing someone new from "home" was a lot greater than I might have predicted. "My name's Fay, from Eugene, and I know both of your brothers, your sisters and several other folks from Lebanon." The easy pace of the moment allowed us to share a friendly reunion and to pal around together for several weeks sharing stories. His family, for one, had been active in the Young Democrats, the same group where I often wrote publicity for submission to local newspapers.

During this time I met a sailor who had been on a battleship in Pearl Harbor during the Japanese attack. He had personally witnessed the death of many buddies and fellow navy men. Our acquaintance evolved around a blunt question about how poorly his clothing fit, not the attack at Pearl. For that reason, I guess, our friendship was close from the beginning. One day, after taking a lot of heat about his clothing, I spotted him near the mess hall wearing a tailored-looking pair of Levi's. Obviously, that wasn't going to go over well with the first officer who noticed him partly out of uniform. It mattered little that the navy issue, called dungarees, was actually a very cheap pair of denim blue jeans.

"Hey, Mac, where'd you get the custom blue jeans?" I said.

"These ain't blue jeans—they're Levis!" He was quite blunt and tried to assure me that pants like his could never be made out of a pair of navy issue.

"You must have a good friend in the states?" I was genuinely curious now.

"I wore Levis all the time I was growing up in Idaho. They're not so special, but just let anyone try to take them away from me." Oddly, no one did.

ISLAND CHAPEL

A typical island church being slowly devoured by the jungle was an emotional sight for many of the men. The curiosity of such a peaceful symbol in a war zone reminded some of different experiences at home as they passed the steepled building on the way to construction sites or back to camp at the end of a long work shift. The quiet building was also a reminder that war was no respecter of places or people; including different races and ages of people. Anyone, anything, and anyplace was vulnerable to destruction.

Every Sunday morning converted natives and sometimes accompanying slaves walked barefoot along the dusty road to the little chapel. Actually it was a small Catholic chapel that had probably been erected by the French colonists at some prewar time. The men of Companies A and B passed the chapel regularly and sometimes, during a relaxed moment, discussed what it meant to them, or perhaps what it reminded them of.

It was a rare treat to see little children following their parents. It was just one of the reminders of the unnatural environment we had created for ourselves. The black natives dressed all in white and were very somber about observing the newly imported religion. The image was a good reminder that all of the world's people were not existing in a living hell. The slightly more relaxed children followed behind their parents, never in front. The distribution of family members also was based on age and the ability to walk unassisted. The sight of little toddlers, especially, was pleasant to both heart and mind. Each of the elderly women always carried a large black umbrella to shield the hot tropic sun and which offer some protection from the frequent rains.

Nearby on the beach was a small pumphouse. It was located at a demarcation between the point where salt water washed over coral formations and an apparent underground spring that provided naturally fresh water. This pumphouse, which was near the chapel, was a popular gathering place for another reason. When approaching this watering hole, one was greeted by tropical, aromatic scents of the most delicate and "romantic" kind. The intensity of the sweet aroma varied with the direction of the ocean breezes and encouraged the mind to visualize a tropical paradise of no worries and immense leisure time spent with loving natives.

"Where're the gardenias?" said one of the older seamen as we walked down a well-established trail for the first time. Just a short distance further and located behind the small pumphouse, large bunches of white, lily-like flowers were growing with abandon, assisted, no doubt, by a small water leak in one of the lines from the pumphouse.

NO SPEEDING

March 9th. Bulletin posted by Lieutenant Tinsley: "All truck and jeep driving must be kept within the speed limit. Unnecessary wear and tear on equipment also increases wear on tires." The notice was odd because the majority of the vehicles have four-wheel drives and are now running in front-wheel drive only to conserve on tires. Also, the posted order appeared exactly seven months after the equipment has been used twenty-four hours a day to meet impossible construction deadlines. Is this guy trying to be promoted to admiral-of-the-seas? Admittedly, at first, most of the drivers didn't know how to handle heavy trucks. The experienced truck drivers were mostly given other jobs. The transportation pool usually consisted of men who lacked skills or maybe just had behavioral problems in their respective companies. A common solution for troublemakers was to transfer such men to the motor pool or to headquarters company and start training them from scratch.

Seven months after we arrived, with half of the equipment worn out, we were given a few truck drivers with experience from before the war and ordered to conserve the rubber on equipment with little potential to last without major part replacements. Working continuously with broken down equipment, we often had little to rely on except the ingenuity of the the Seabee motto: "Can Do." The motto could sometimes be modified to include "Work, Work, Work."

The message that seemed to come down from the top brass was something like, "Well, you men are living on an island paradise in the South Pacific. If it were not for the war effort, the Navy would have built its fleet recreational park here." Sure, a recreational area built by Seabees but off-limits to the construction battalion men who weren't officers. However, there were moments for levity on this "island paradise." One ship's captain, for example, required his men to wear full whites ashore for two hours of beer drinking and playing baseball. Once the men had their fill to drink and played aggressive baseball by our "island rules," the whites looked as if they'd never come clean again.

HURRY UP AND WAIT

The play and good-clean-fun was soon over and the orders went out to do another "RUSH" job that required the men to work seven days a week and to work long shifts around the clock to complete some particular project. Oncc done, the men could sit idle for a week to ten days and then the same cycle would start again. The men's morale tended to fluctuate from extreme highs to extreme lows depending on the war news, the reasonableness of work scheduling and the quality, if any, of the tools and equipment available to do the job. First class carpenters, for example, who possessed good civilian experience were often obliged to work with missing, inadequate, or rusty tools. Eventually, it appeared that the men weren't concerned about finishing jobs "on schedule," or even finishing the job. During such down times the 8 or 12 hour shift crept along at a very slow pace. Each man knew he was capable of doing much more, and preferred to, but the team solidarity slowly crumbled until jarred by the next real war crisis.

There's a big dose of military bull-headedness in being forced to wake up at 5 am each day when the work shift doesn't begin until 7:30 am. Under the circumstances, it's not so much a question of military order and discipline as it is adapting to bizarre living conditions. Men who do manual labor all day need good rest before doing more of the same day after day. In the monsoon driven tropics, however, there's little rest in a humid, soggy cot with an uncomfortable mattress. Often we come in soaked to the skin, hang our wet clothes up, and find them just as wet in the morning and slip into wet shoes. We'd have jungle rot for sure if it wasn't for the heat which actually dried us out in about thirty minutes once the sunlight is warm enough. The poor sleep, long hours, soggy clothes and poor equipment are not relieved by the food. Too often mess hall food was a "mess" in every sense of the word.

KEEP A LID ON IT

It seems that maybe the common purpose of winning the war and getting it over with has dampened fights among the men or other misbehaviors such as stealing, faking illness, or just going berserk. Thoughts of insubordination are quietly said in jest or not said at all because of the potential for being hauled to the guardhouse for sedition or other forms of high treason during very serious war times. Restraint and continuing to do what could be done only under very difficult circumstances was done because of "patriotism" and the navy's ironclad rule of "do it or else." In stark contrast, there were many instances of civilians doing comparable work in the tropics for very high salaries and benefits that should last a man a lifetime. Even with such incentives, they often gave up and returned to the States. The moneyed-class, the French and English plantation owners, simply move their wealth to safe havens in Australia, New Zealand, etc., and wait out the war while enjoying a period of physical and mental recuperation.

Here we are, metaphors for the slaves owned by black natives of the island as well as by plantation owners, "fighting" to save our homeland from a mythical invasion. We're virtually sitting ducks for the strong Japanese military bases located only a few hundred miles away. Oddly, the enemy could send only one bomber over at a time and rarely did it do any real damage. We struggled daily with hard labor and gross discomforts to eke out a little progress in the "total war" effort. The commitment to "country, family and apple pie" kept us going despite the feeling of being kicked around by mostly indifferent forces.

The impression of the moment is that the base commander prefers to have his island in a state of turmoil rather than a smoothly running machine. The various department heads reporting directly to him continue to allow the destruction of mobile equipment and the waste or absence of essential equipment and supplies. Most people would readily understand that on an island virtually everything has to be brought in, cannot be created out of nothing. It's not the enlisted man who can see to the materials that are critical to have on hand. It's such a quagmire at times that it's not surprising that they don't allow the battalion to return to the states and share their stories with the raw recruits who are being pumped up to go out and save their nation. "Join the Navy. See the world," etc.

February 10. Company morale in general has been running quite low. We never quite knew when the feelings may be symptoms of malaria or some other dread tropical disease. I requested permission to check with sick bay, but one of the arrogant ensigns prevented me from doing so. It turned out to be ordinary dysentery and passed after maybe a dozen trips to the head. It's extremely easy to become dehydrated, under the circumstances, and it takes several days to make a comeback. The latest monsoon hit so hard there hasn't been any real work for several days. I even had a little extra time to reflect on who might have stolen my last bottle of Coke. Might have been one the officers who wanted something to mix with his bootleg hooch. A few people around here have about as much character as the black rats who made it ashore from sinking ships centuries before.

MORE OF THE SAME

April 4th. A full month, mostly filled with rain, has just passed. A miniature flood continues to run through the camp. The *USS Monroe* docked in the harbor today. The continuous heavy rain causes a lot of dissention among the troops. Men get touchier each day and irritable with the company commander as well as the base C.O. Eight months of unimagined austerity and brute labor in the tropics is more than too much. The majority of the camp area is one big mud slick as it's located at the base of a hill and gets maximum runoff.

One of the interesting parts about this unnamed hill is the army unit located at its modest summit. The army guys have their own PX and usually have a different film to watch every night, about the only good life to be had around here. It's a ten-minute walk up the hill to visit with the guys who survived the sinking of their ship in the Segun Channel. It was one of the quirky events that could have worked out completely differently. Apparently, the ship's captain didn't want to wait for the pilot boat to take him through a known mine field. The bull-headed captain lost it all. Luckily, most of the men were topside looking at our island when the mine exploded. Many were able to swim to shore and all the ships in the harbor sent small boats alongside to pick up the rest of the men. The cargo and the ship itself was a complete loss. The ship had been loaded with soldiers and supplies intended to provide relief for the hard pressed marines on Guadalcanal. Now, very stupidly, the men were stranded on Espirito Island and dependent on borrowed supplies for everything they needed. However, no work assignments were handed out. It was one of those slow times when extra laborers weren't really needed. So it goes.

MOON SHINING THROUGH THE PALM TREES

Full moon tonight. The kind of night old Tojo probably dreams of. However, just in case Tojo's boys send another submarine offshore to lob in a few more shells, they'll have a little reception committee. A detachment from Company B just installed an 11 inch gun on one of the off shore islands. A full moon could be just the kind of night needed to catch the silhouette of an off-shore, enemy sub.

The evening weather, when it wasn't raining, could be very pleasant. We watched the moon rise while sitting in an open air theater we had created with 1" x 12" planks. Rising out of the near total darkness from behind a far away island, the moon reflected across the waters of the channel. At certain angles it shone through the graceful palm trees for a period of several hours each night. The peaceful, romantic images and almost complete silence, except for the mild surf and breezes, reminded me that this war-torn world had far more to offer than rumbles of big guns and multicolored flares crossing the horizon. The movie tonight, *Take a Letter, Darling*, reminded me of my home state when they incidentally included scenes of dancing, walking in the woods, and horseback riding. In my former life, before my expense-paid trip to the tropical South Pacific, I had spent many pleasant hours dancing and riding horseback.

ISLAND GAMES

Part of our company was moving to another island. I was uncertain, but I thought I might go with them soon. Few were surprised to learn that they had been reassigned to different locations with little notice to do small construction jobs for the admiral's people. The transient nature of the work and living assignments meant entertainment was whatever a bunch of guys could create for themselves. Some devoted serious attention to brewing and caring for wine made out of potato peelings, or raisins, or whatever might be available. Occasionally, out of shear boredom, guys either raced or chased each other from tent to tent, tree to tree, or truck to truck. Some played ball games or cards, and there was the periodic brawl. Last night I witnessed another of Company B's favorite sporting events, R. B. Davis versus everyone. He didn't last long. Too much wine on that particular occasion. So it goes.

R. B. had been edgy and antagonizing for several weeks. He was a big enough guy to be respected just for his size. It wasn't clear why he seemed to have such a touchy attitude. Probably something that started back in the States and never got resolved. He has been a locally famous tough guy but was now down for the count. He managed to fall off an elevated area during his show brawl and broke his leg. After that, he was crankier than ever to be near.

Another guy, Leo Hatch from Company A, had the misfortune of having to share hospital space with R. B. Little doubt that he's looking forward to any chance he gets to leave this island—and R. B. Actually, Leo and I have been buddies since our shipping out together from the Union Station in Portland nearly a year and a half earlier. I'd really miss him, but some guys simply need to get out of these islands if their health is ever going to improve. Sometimes it's the mental health that also needs a change of venue.

Easter Sunday, April 25th. No Easter egg hunt this year. It wouldn't be very military, maybe not masculine enough in a war zone where there were shortages of so many things and little chance to duplicate the world we left behind. Believe it or not, last year in Portland, at the Columbia Hunt Club's Annual Easter Egg Ride and Hunt, I came in second. What fun at the time. This Easter I'm quietly reading a book and munching on two pork chops (mutton, I think) and several pieces of toast taken out of the mess hall. Oddly, toast was a special treat as it was seldom served. The day was a leisurely one. No monsoon to contend with. Was able to do a little trading with one of the natives who came through the camp offering bracelets for almost anything edible. Trying to decide if it's worth it to walk up the hill to the army PX, or maybe catch a truck ride to the airfield and just hang out.

Friday, April 30th. I've been working quite intensely for the past week and have felt okay. After lunch I was unable to keep the afternoon meal down. It came on quickly and I knew I was actually quite sick and checked into the sick bay. I discovered that Leo Hatch was about to be shipped to New Zealand because his own health was not improving. The hospital people have it good. That is, when you're physically up to enjoying it. You can get a cold bottle of Coke for only a nickle. Tastes just like home. A little rum, if we only had some, would make a very satisfying, tropical drink.

FRIENDLY REMINDERS

May 2nd. On this date I was walking past the pumphouse and the small chapel when I saw a couple of sailors coming toward me. Not an unusual event but there was something familiar about one of the guys almost immediately. There was something distinctive about the way one of them was walking. After a few moments I began thinking to myself, "Hey, that's Fritz from the old home town."
As we got closer I said, "Are you from Oregon?"

Fritz seemed to recognize me immediately. "Hey, Fay, what you doing down here?"

Fritz and I had gone through junior high and high school together. Right after graduation Fritz enlisted in the navy and simply vanished for four years. It was at the end of his enlistment that I saw him again. He was casually thinking about re-enlisting at the time but nothing definite. Now, we're both many thousands of miles away, both in the navy and sharing the same bit of space on a tropical island in the South Pacific. I remember now that he was stationed aboard a small service vessel. I was able to visit him once, but their ship was small and the C.O. didn't like visitors aboard so, we were barely able to talk at all. I stopped back the next day and discovered the vessel had pulled anchor and steamed away.

Have been working ten hour days and seven days a week for some time now. The weather had been much drier and it was more comfortable to sleep at night. Another reason, the tents were screened now and we had the unbelievable luxury of screen doors. It meant no more lizards or other crawly things sharing our beds quite so readily. Each man made his own, custom chair at one time or another. I'd been finishing a hinged desk to serve as a night table and place to write. As time allowed, we made various improvements of the tents to protect from the monsoons. We sometimes boasted of living in a four-room tent. Each time a new tent was issued, we just added it to the leaky one.

Intensive efforts are being made to haul lumber from the New Zealander's camp to our camp to be cut to specific sizes for the admiral's headquarters buildings. On one of those trips, we picked up a couple of sailors. At the time they were off their ship with instructions to obtain certain construction materials, if possible. After talking with both guys for a while, I learned that one of them had gone to the University of Oregon and lived less than forty miles from my own home. Their requests were reasonable enough, and being from the old home town didn't impede their request. We made arrangements for them to meet us at a designated place and time and we would haul the request lumber for them. In return we were invited aboard ship and treated to some *good* meals.

The ship boarding turned out to be the food event of the year. It was almost the Fourth of July, and extraordinary evening meals were being served up. Don Swamger, C.M.2/c introduced me to the chief in charge of the mess haul. He also described how the Seabees had made special efforts to haul the requested lumber for their ship and that I was from the same state. Without a doubt, the chief was really one swell fellow. He showed us the menu he had prepared for the Fourth of July—it was really something—and he invited me aboard to share the festive meal and join them at their scheduled baseball game after dinner. Following the most extraordinary holiday dinner I'd have in at least two years, we went ashore to the Fleet Recreational Park. The beer I didn't need I gave to my newly acquired friend from Oregon. He could put them away like the legendary "drunken sailor."

REASSIGNMENT

Several weeks later, following a period of good weather and increasing temperatures, the lieutenant asked me if I would go with a detachment to a small, off-shore island and assist Seaman McLaughlin, 3/c Cook, in providing meals for 24 men. I agreed to go but insisted that it not be as a mess cook. The change of pace and a different bit of jungle to look at was a welcomed idea.

Before leaving, I attended communion and church services at the island's small chapel. I bought up a supply of candy for possible trades with natives on the new island and had one great dinner aboard a ship known only as AV 8. It had been one year exactly since I boarded the *USS Monroe* at the docks in San Francisco and began my personal odyssey in this South Pacific "paradise."

The new island did have a distinct change of pace and could almost be described as a picnic compared to the intensive labor and long hours on Espirito. Here, helping to cook for 24 men plus McLaughlin and myself was good duty. Our first evening meal was pork tenderloin (lamb?), fresh potatoes, green beans and applesauce, and plenty of it. Sometimes it's great to be immediately appreciated for doing a job that only remotely contributes to the war effort.

Here, we were camped in temporary quarters on the exposed ocean side of a small island perhaps three miles long and half a mile wide (if you're a crow or bird of paradise). One hundred yards into a dense tropical jungle can appear to be all of one-half a mile. It was quite small but not uninhabited. There was one plantation and at least one band of natives living there. The French planter and family had been removed to another location. Rumors were about that in revenge he had left a "head hunting savage" in charge of the plantation to keep other natives from making any raids on the grower's personal residence. The plantation itself consisted of a few acres of coconuts and a small herd of beef. The island's only food crop, except for a small garden, was coconuts and bananas. There was also evidence that the natives had semi-domesticated the wild pigs and cultivated tropical plants to supplement their diet.

When we first arrived on the small, and I guess unnamed, island, we found a U.S. colored detachment from the coast artillery that had been stationed on Espirito. They were manning two moveable field rifles and some powerful lights. Once settled in, it was from their separate camp that we were able to receive the war news each night and some fresh beef. Protocol prevented us from joining the two camps together.

For an outpost in an incredibly remote spot on the ocean, our compound was quite well stocked. We had a portable still for fresh water, showers, and laundry, but not for making hooch. There was a generator for light in our tents rather than kerosene lanterns, and the galley had a four-foot by six-foot walk-in cooler—a convenience that was indispensable for keeping food in the tropics. We had a mess hall tent, a headquarters tent where the ensign, corpsman, McLaughlin and I slept, plus two or three tents for the other men. The supply boats were scheduled for deliveries twice a week—but—the Can Do Seabees had their own channel bus service to make life just bit more liveable.

INTER-ISLAND TRANSPORT

Earlier, two mechanics had befriended the marine lieutenant from Colonel Carlson's base camp. The mechanics traded some know-how for a couple of outboard motors and a twenty foot wooden boat. Of necessity they travelled between our camp and the main island once each day. Our Seabee pipeline.

Our equipment at the small, unnamed island consisted of an HD 14 heavy bulldozer, one dump truck and a ten-wheel troop truck. Our only self-defense was our bare hands and the fire power of the lieutenant's .45 caliber side arm. Real tough to say how long we might have held out if Tojo had sent a few battalions of his boys to take this island back. For the moment we camped in a small clearing inshore from the only sandy beach on that side of the island. It was sometimes difficult to visualize this small patch of ground as a war zone. When the tide was in, the swimming was perfect, regardless of the skimpy size of the sandy beach. It was a small, crescent-shaped beach about twenty feet in diameter. For recreation we could ride the swells out and back over the highly dangerous coral without real danger of scraping the ragged and sometimes poisonous areas that characterize the coral reefs. The reefs were carefully checked to rule out the possibility that dangerous sharks might enter that small space.

Seaman Meadows was a part owner of the inter-island boat and motor. In the evenings, when the tide was high and the water smooth, he would take fellows for a ride over the briny deep. The trip, however, was not always without incident. One evening, for example, when all was going well, he invited me aboard. The opportunity for a change of scene from this very small island was welcomed. We were barely offshore when the swells began to roll up to four feet and higher. I tied a line to the motor and the other end to a bucket. There were six of us in the small wooden boat—the ensign and the others. We were soon heading straight into serious breakers. Soon after reaching the cap of the wave's crest, the small boat dropped like a rock. After several hard breakers, the boat began to turn on the broadside to the caps. The first wave over the transom drowned the motor. With the second wave we were parallel with the breaking rolling wave and in the process of being capsized. As the little boat rode the swell up, the ensign yelled, "Jump!" As he and the men jumped the swell dropped and so did the ensign. He landed with a hard belly flop into the water. I jumped clear but was on the ocean side of the breaker. The ensign and another man floated ashore on the swell but I was being dragged out to sea because I was still holding onto the line attached to the motor. I did everything I physically could to pull myself up to the overturned boat and hang on. I realized I was out of sight of the shore and beginning to feel a strong sense of panic. The suddenly strong winds only carried my shouts further out to sea. Soon I was a half mile off shore and expecting to become fish bait or an unarmed, one-man assault upon the Japanese Empire. I tried to signal with my arms in the assumed direction of the shore and busied myself trying to right the wood boat and load the up-turned motor back into position. It was close but in the rough weather and without leverage to flip it over, I couldn't right the boat by myself.

Throughout this brief interval, the undertow was pulling boat, motor and me out to sea as I continued to hold onto the line attached to the motor. I had been dragged at least three quarters of a mile and the swells began to die down. I was able to watch the wave action carefully and prepare for another seven-wave sequence. I was beginning to feel that I could take care of myself and wasn't going to die because of some rough weather. I had spent enough time swimming in milder ocean swells to feel I could survive as long as the boat didn't go under. I wanted to do everything possible to salvage boat and motor and live the Seabee motto: Can Do. As the abrupt winds began to die down almost as suddenly as they occurred, I began to think of the experience as high adventure. Like a small and predictable roller-coaster, I felt the excitement of every breaking wave movement and wanted to prolong the experience. My real concern was whether or not I could save the boat and motor by myself.

Drifting outside the severe rolling action of the breakers, the ocean was actually becoming smooth. The current appeared to reverse, and I could see that the capsized boat, motor and I were being carried toward the shore. The only bad part was the current was taking me toward the unknown beach on the lower part of the island. By keeping the motor heading on the side of the capsized boat, the swells turned to brakers and carried me and the boat up on a beach about three quarters of a mile down island.

I was privately overjoyed to be alive and ashore without serious injury. The emotional boost of the unexpected adventure quickly wore off when I realized the boat was unrepairable. The motor was ruined. I beached the vessel as well as I could above the high tide mark and began trudging back to camp to surprise the guys who made it ashore and probably thought I was lost at sea.

Shortly after the trauma of losing the wood boat and and outboard motor, there was another brief trauma after the evening meal when we spotted an unidentified vessel on the horizon. It was traveling down island from us and resembled a surfaced submarine. Nervous tension was building as several minutes of speculation didn't help to relieve our suspicions. If it was an enemy sub, did she see our camp? Finally, thoughts of seeking cover and of defending ourselves with one .45 caliber handgun were relieved when it became apparent that it was a sea-going barge towing a high crane. The mind can quickly play serious tricks when the outcome could very well be the difference between life and death.

BOILED LEATHER

About one week later, on a clear night, the Army Light Unit came to our camp with a dressed beef. They offered to give us half a beef if they could use our cooler. Without hesitation we agreed. The following evening, McLaughlin and I fried real red beef steaks for the men. They turned out so tough it was a real effort to eat one, but the meal was worth the effort. The following day we began boiling the beef, but the results were about the same. It was so bad, in fact, that we preferred to go without. When the army unit returned some days later we said, "Next time, shoot a calf!"

This little incident built up our desire for tender, fresh beef to the point that several of us were delegated to go on a hunt. The plan was to borrow the ensign's .45 and for Meadows to drive the ten-wheeler. We located the semi-wild herd in a relatively open area and gently rolled toward them. We discovered one cow down. The female cow had been wounded, apparently by the soldiers and was injured enough to be slowly starving to death. The ground around her front legs was a patch of bare earth, evidence of her unsuccessful efforts to get up. The pathetic sight made me feel ill and disgusted. I slowly approached her from the rear and put one shot in the back of her skull. The effort to get close to an almost wild calf was a dismal failure. My aim with a .45 caliber handgun was not improved by the bouncing ten-wheeler. Still, it didn't help that Meadows kept making cracks about my obviously poor marksmanship.

The word did get around about our unhappy experiences. To our surprise the army unit reappeared some days later with half a calf. The men in our camp were crazy about being able to eat all of the steak they could possibly put away. We even had steak again the next day for breakfast and lunch and enjoyed every bite. Because of our electric cooler, we were able to wash it down with ice cold coconut milk.

SOUVENIR COLLECTING

July 18th. I hiked down the coral studded beach today. It was one of those rare opportunities when I had some completely unstructured time just to walk and daydream without some issue to resolve. I set about looking for the most colorful coral snakes and shells I could find. I did see lots of snakes but avoided disturbing them in any way and maintained a safe distance because of their reputation for lethal poison. Along the way, I found an especially attractive shellfish. It was definitely a trophy-quality specimen, even if I couldn't identify it, so I carried it back on my return journey. It was equally uncertain if this particular shell fish was edible. Instead, it was soaked in fresh water for roughly one week. That allowed the body to liquify and the shell to be washed clean and saved.

Among the other souvenirs collected as one of our few non-work-eat-sleep activities was trading with the natives for whatever interested us at the moment. Many of the guys would give cash for boar's tusks. I recall paying a native up to thirty-five dollars for one distinctive pair. Sometimes I was able to make trades with canned Spam, canned bread, or canned salmon. The latter was a distinct treat for the natives and helped a bit to ante up the trades. Having so little to work with and so little cash to make outright purchases themselves, the black natives were sharp traders.

The native chief of this small island had a mostly complete uniform of several different nationalities from the various different services. Actually, he had one, more or less complete outfit, for each day of the week. In spite of very high temperatures, he was seen wearing a different uniform each day and displayed immense satisfaction with the image and the powers his new appearance gave him with his own people. For pure enjoyment, the chief definitely got the best of the trading deals.

Occasionally there were little bonuses from our own people. The PX, for example, handed out free cokes and beer today to share in the profits they had made to date. A couple of men were able to pilfer a couple of live grenades and went fishing in the tiny bay. It was a real treat when they returned with an assortment of fresh fish to spice up our traditional menu of Spam, beef, and pork (mutton?).

PARADISIO

The few quiet moments are the time to enjoy the virtual ecstasy of the tropics. The days and nights are so similar and mostly pleasant that it's difficult to keep track of one day to the next—or to care. The view in every direction was stunning and showed little sign, or even potential for, human development. The warm wind, for instance, blew constantly ashore from the ocean. The wave action and the surf action was an unending rhythm disrupted by the occasional boat that provideed inter-island transportation in predictable patterns and served our little compound. The absolute brightness of stars and the full moon reflecting on a blackened ocean was more dramatic than any postcard could ever capture. The temperature of the ocean water was warm and inviting except for the known risks of jagged coral, the occasional strong current, or predatory sharks.

August, 1943. I had been on the islands of Espiritos Santos for one year now.
Life had a much more pleasant pace about it than it did when I first arrived. The work load was not so intensive and there were opportunities for simple pleasures, such as swims in the ocean. Even in our protected swimming area there were sufficient rollers and breakers to be a real challenge to a swimmer. No one thought of it at the time, but I doubt that the waves would have been sufficient for surfing except during a dangerous storm. Also, there was the sheltered cove at low tide for the non-swimmers. The high tide allowed the good swimmers to be well above the dangerous coral reefs.

Objectively, life, under our preset circumstances, was quite one-sided. There was the obvious absence of any female contact or companionship other than the occasional glimpse of a black, native woman. The common stories about revengeful head-hunters were more than enough to prevent even the most lonesome man from showing any social interest in a native female. Nor was there any prostitution, or loud, smokey honky-tonks as might exist in other parts of the world. Our social time and recreation was whatever could be done with "virtually nothing" as young boys do. Because of our small numbers and extreme isolation, the navy simply made no effort to provide recreation for this detachment.

Each evening those who wanted could take the ten-wheeler and drive the three very bumpy miles to the army camp. There, we could listen to the day's war news, anything from the outside world and all the music we could get in. Swapping inter-service lies and other stories about home, girls, work, and girls was great sport. The army lieutenant had invented his own competitive sport. As the only possessor of a .45 caliber handgun on the island, he amused himself by shooting rats in his tent. When he announced what he was about to do, the rest of the men gave him a wide berth to avoid being hit by stray fire.

For some of the guys, including me, one of the main purposes in life besides work, work, work was "to live to eat." The first rule I had learned from boot camp was if you wanted to eat better, work in the galley. Good, individual foods had a way of becoming nondescript mush after being combined and allowed to cook in large portions for long periods of time. A galley helper, in contrast, could fix up individual portions and cook it just enough to satisfy one's personal taste, not like the nondescript substance dipped out of ten-gallon or thirty-gallon cooking pots.

A rumor circulated that we might soon be shipped back to the Great Lakes Boot Camp to help in training new recruits. My work detail suddenly changed to KP and assignment to the gear locker. Orders were received to clean up and store all available equipment. There was a real sense of optimism and relief that we might soon be "re-entering the world." I recall looking at a note I'd made to myself during a down moment as follows: "Too much navy, too many regulations, too much nonsense. Too many of the few perks (whiskey, etc.) were going to the gold braids. Wish to hell I was out of here." Many times I lay on my cot, unable to sleep, wondering what kept me from cracking up, or the other guys from doing the same. The mental strain of thirteen months of almost unrelieved labor, periodic illness, poor food and worse sleep eventually rips the lining out of any civilized kind of existence. A few guys did crack up and had to be sent back to the States.

Keeping one's sanity under the circumstances and the external threat of being either killed, wounded, or captured before the end of the next day had a lot to do with pacing. That is, we took each twenty-four hour day gently as it came, and reasonably paced each activity without a sense of panic or major hurry. I looked forward to each of the opportunities to do the little things that were personally satisfying to me—to read, to write, to do something for fun, to enjoy the buddy time together, to enjoy the moment and not try to live in the past or future. Why kill ourselves to complete every job in one day when there will be a tomorrow? We learned that life was pretty much the same, day in and day out, but it was possible to make things happen. It was finding the little and often trivial things to be excited about, or committed to, that made that remote, tropical life more meaningful and tolerable.

Monday, August 16th. Another monsoon today. It rained so hard the water filled my shoes. The ensign was complaining of a chill in the air as the seasons (and reality) were reversed there. Still, daily instances of rushing to get somewhere and then having to wait idly remained the same. The winds are strong enough to put serious whitecaps on the breakers hitting the beach. The latest news was that S.C.A.T. (South Pacific Air Transport) moved their headquarters between our present camp and the admiral's headquarters on the larger island. Probably to protect one man.

Planes are arriving daily from Pearl Harbor and carry more detailed bits of news as well as various types of merchandise that we can't get in any other way. The lieutenant in charge, for instance, wanted an island telephone. The island communications unit informed him that there was no wire available for new phone installations. Even if he had all of the equipment, they were reluctant to make hard-wired connections because of the hazards of stringing wire in the jungle. The lieutenant, however, was determined and detailed a unit of men to remove a mile of wire from another source and rerun the wire where he wanted it so he could have a personal telephone service—with a one mile range. So it goes. Personally, I'm drinking a lot of fruit juices, have a good tan, and, like most guys, hope to be home by Christmas.

A. Clark Fay, Oct.,1943, Jan., 1995.

A local native man with Company "B" in the background.

South Pacific native huts. Here, located only a few yards from Company "B," these natives received many items of clothing and food (including cigarettes and candy) courtesy of the Seabees. Also, Seventh Day Adventist missionaries had lived on this island for many years and had acquainted the natives with many religious holidays. It was common, for example, for natives to weave roof panels from coco fronds and to sing gospel songs at the same time. In contrast, there were stories of hostile "Bushmen" with a history of cannibalism and a male-only society that was sustained by stealing young boys from various tribes. All Seabees were warned to stay out of the jungle unless fully armed. Often, "friendly Bushmen" came out of the jungle with ripe bananas, pineapples and handwoven goods to make trades.

View of Company "B" from offshore. Up to four tents were combined to protect against the heavy rains. Each tent was eventually screened, had board floors and electric lighting. The camp was equipped with a movie screen and fresh water showers and attracted many visitors from the Marine camps as well.

Christmas, 1942. Various Seabees are here sharing candy and trade goods with island natives. The lifestyle of the natives appeared to be the simplest possible—basic existence under the sun, moon and stars. Food was easily obtained from the soil, banana trees, coconut trees and from the ocean.

Another coconut grove. Each of us has probably used hand soap with coconut oils that originated on a tree in the tropics.

A Seventh Battalion work crew. Dress of the day varied with the weather. In overcast conditions men often went shirtless. The intense sunlight at other times changed everyone's "natural" body color.

Navy Advance Base Shower. This was the first *hot* water showers on Espiritos Santos in 1942. The hot water was furnished by a gasoline-driven "still." All military personnel were eager to use the showers when possible; including the crews from the 98th Bomber Group returning from Guadalcanal.

Company "B" men relaxing here at the end of a tropical day (late 1942 or 1943). It didn't rain all of the time. Many hours were spent enjoying the warm days. Pictured here are Everard, Klama, Kitchens, maybe Kopp, Brown and a member of the 2nd Marine Raiders.

American burial site of Japanese war dead on the island of New Hebrides, 1942. The fencing was erected to protect the cemetery area.

Recreation hall for Companies "A" and "B" located on Espirito Santos. The roofing provided reasonable protection from the monsoon rains and the burning sunlight. It contained assorted books and magazines from home and the ever present ping pong table.

When working on a detail or just hiking through the coconut groves there was always an abundant supply of coconut meat and liquid available. The large, dry, gray ones on the ground can be broken open and the snow white meat can be eaten without further preparation. If thirsty, pick a green one can carefully open one end and the liquid inside makes a delicious drink.

The island of New Hebrides was the land of coconuts. Always reaching for the stars, the coconut tree furnished fresh fruit and warm coconut milk.

Another work day of Island "X." This site was also known as "Strawberry" Island and was the Navy's southernmost advance base in the South Pacific. Pictured are Everard, Brown and a man from the 2nd Marine Raiders (Carlson's Raiders).

American Cemetery. One of many final resting places for hundreds of service men who gave their lives for their country at Espirito Santos, New Hebrides, and others from 1942 onward.

Rene' River, Espiritos Santos, was often filled with servicemen having their first fresh water swim in months. The river was a source of great relaxation from 1942 on.

UNITED STATES NAVY MEMORIAL & VISITORS CENTER

Bronze relief sculpture: **"Seabees Can Do"** (West Wall Sculpture #1)
by Leo C. Irrera, a WW II Seabee.

Sponsors: U.S. Navy Seabees (former and current active duty), Navy Civil Engineering Corps officers, and related organizations. U.S. Navy Memorial Foundation, 701 Pennsylvania Ave., NW, Suite 123, Washington, DC 20004 Tel: (202) 737-2300 Hours: Mon-Sat 9:30 am to 5:00 pm and Sunday Noon to 5:00 pm.

(J. Kimmel photo)

SEABEE

HENRY B. LENT

New York: The MacMillan Co. © 1944 (Edited version reprinted by permission)

ACKNOWLEDGEMENT

Time and again, among the first American detachments to land on enemy terrain have been the "Fighting Seabees"—the heroic volunteers who make up the U.S. Navy's famed Construction Battalions.

Armed with tommy guns, carbines and hand grenades, they have helped the Marines clear one beachhead after another so that their landing barges could bring bulldozers, giant power shovels, scrapers, and other equipment for building bases and airfields.

The Seabee motto reads, simply, "*We build—we fight.*" But the men themselves have another motto that goes like this: "*Can do, will do—did!*"

They take particular pride in telling about the time a company of Marines was swarming from landing barges onto a Japanese atoll. As the Leathernecks plunged up the beach with fixed bayonets, a Navy lieutenant calmly strolled down to meet them. Holding out his hand, he said, "The Seabees are always happy to welcome the Marines!"

The Seabee battalions are scattered all over the world. From the Aleutians to the Solomons, and from Italy to Iceland, you'll find them running their winches, their pile drivers, their "cat" bulldozers—building airfields, barracks, bridges, docks, hospitals, storage tanks for water and fuel, or loading and unloading ships, often under fire.

One Seabee, based on a certain "Island X" in the South Pacific, described the Construction Battalions this way: "We're the newest outfit in the Navy, mister—and the toughest. But if you're looking for glamour you'd better head for the quarters of the Navy aviators. We can tell you where those quarters are because we built them. They're all over by the air strip we slashed out of the jungle when we weren't ducking into trenches to man anti-aircraft guns. But when it comes to glamour we've got about as much as a bunch of warty toads squatting on a rock. We've got no wings on our chest—just hair. Our uniform is soggy boots, dungarees, and a rain hat. When the Japanese pester us, we kill 'em and get on with the job."

And that, without exception, is the kind of spirit I found among the Seabees who are in training at Camp Endicott, in Davisville, Rhode Island. Up to July, 1943, all of them were volunteers. Many gave up good jobs in civilian life to offer their skilled services to the Navy. They ask no favors and expect none. Their one thought is, "Let's win the war and get it over with!"

To the many officers and enlisted men at Camp Endicott who helped me write this account of Seabee training, I am deeply grateful. Especially to Lieutenant Thomas Bradford, aide to Captain Frederick Rogers . . . to Lieutenants Hughes and McLaughlin, who explained the many phases of a Seabee's technical training . . . to Lieutenant Rogers, instructor in military training . . . and to Chief Petty Officer Vaughn, who, among other things, put an eighteen-ton "cat" through its paces for my benefit.

Many of the incidents in the latter part of this book were taken from letters written home from "Island X" in the South Pacific by Seabee L. R. R. Martinez, Pho.M. 1/c. My sincere thanks to him for the privilege of reading his interesting letters and using them in this way.

My thanks, also, to the Navy Department for the privilege of visiting Camp Endicott, and for the official Navy photographs with which this books is illustrated. Henry B. Lent

HERE COME THE BOOTS

The long train of coaches jolted to a stop on the railway siding. Bill Scott looked out of the dust-covered window. What he saw resembled an ordinary town, with straight paved streets and cross streets laid out in a regular pattern. But, instead of stores and all sorts of houses, there were only long rows of olive-green barracks and metal Quonset huts with their curved roofs.

"I guess this is the end of the line," he said, turning to Tex Rogers, the boy who had shared his seat in the coach on the trip down to Davisville.

"Yep, this is all right," Tex drawled.

They both picked up the small bags in which they had packed the few earthly belongings they had been permitted to bring with them to the Naval Training Station. The aisle was already crowded, as the new recruits jammed their way toward the door at the end of the car—all talking at once and in loud, excited voices.

"Let's take it easy, Bill," Tex suggested. "We'll be a long time here at the base. Why should we risk breaking an arm just to get ashore a couple of minutes sooner?"

"All right, gob!" Bill chuckled. "We'll wait until the big push is over."

Bill liked Tex Rogers, even though he had known him only a few hours. He was amused at the way Tex, who had never even been in a row-boat before, was already trying to talk like a sailor—going "ashore" from a train. Tex was a rancher's son from down near Abilene, Texas. He still wore his high-heeled cowboy boots and broad-brimmed hat. Until he came North to join the Navy, he thought *everybody* dressed that way.

Just before Tex left home he had come across a copy of an old *Bluejacket's Manual,* which he had studied all the way on the trip. He already knew that in the Navy you call the floor of a building the "deck," and a wall a "bulkhead," just as on a ship.

He had taken a lot of good-natured kidding about his cowboy boots and hat, but he didn't care. Some of the other boys were wearing clothes that seemed just as funny to him. Some had sweaters instead of coats. There were plenty of old battered hats in the outfit, too. Many of the boys were wearing suits that had seen better days. Quite a few were in their shirt sleeves, without neckties. But that was all right. The recruiting officer had told them, "Travel light. You'll have to send all your civilian gear home as soon as you get there. The less you take, the less you'll have to send back."

Shifting from one foot to another, the recruits filed slowly out of the railroad coaches. There were about five hundred of them.

Several of the boys with whom Bill had become acquainted on the trip down came over with their bags and suitcases to the spot where Bill and Tex were standing. There was Jim ("Slim") Brown, a garage mechanic from Connecticut...Nick Torrio, whose father ran a trucking business in a small New Jersey town...Dick ("Swede") Swanson, from Minnesota...and Terry O'Rourke, whose ambition in life was to become as good a steel-construction engineer as his father was.

Now a chief petty officer was forming the boys into a long line, two abreast. Bags in hand the recruits started down the main street toward the Receiving Barracks.

Overhead a formation of fighter planes from the near-by Quonset Naval Air Station streaked low across the sky. No sooner had they shot out of sight over the trees than a big twin-engine

Catalina patrol bomber zoomed by, coming slowly for a landing on the bay.

Two sailors in a Navy Jeep jerked to a full stop at a cross-street intersection while the new recruits tramped by.

"Here come the Boots," the driver said to the other sailor.

"Yeah," the second Seabee replied, "They look sort of raunchy, don't they?"

"Maybe so," the first gob shot back. "But wait until you see them a few weeks from now! I guess we all looked something like that when we first came aboard the Station."

As the last of the recruits passed by, the driver shifted into gear and the Jeep continued on its way.

When the boys reached the Receiving Barracks, a yeoman wrote down each one's name on his list. Then they lined up to receive their "night issue," which consisted of a mattress, pillow, blanket, towel, and a bar of soap.

Each recruit was assigned to the barracks which would be his living quarters during his boot training. Bill Scott, Tex Rogers, and Slim Brown drew the same barracks—number 28. Loaded down with their bags and night issue, they trudged down the long street to their quarters.

"Look! Just like the Ritz!" exclaimed Bill, throwing his bag and other gear down on one of the hundred bunks that lined the long bare dormitory in Barracks 28.

"It sure is." Slim chimed in. "A radio in every room, and a maid to turn down our blankets for us at night. What luxury."

"It may not be luxury," Bill admitted, "but here's one tired gob who's going to turn in."

And he started to make up his bunk.

"Me too." Tex said. "We've got to hit the deck bright and early tomorrow morning."

But long after "Lights Out," Bill was still wide awake. He was too excited to sleep. From some of the other bunks came a strange chorus of snores. Quite a few of the boys had been traveling for days and were dead tired. Bill smiled as he lay there in the dark, listening. He wondered if the rest of them felt the same thrill he did on arriving at the training base.

"I'm a Seabee—in Uncle Sam's Navy," he thought to himself proudly.

This was the day he had been waiting for—the chance he wanted, to show that he had the stuff to become a "Fighting Seabee" in the grandest Navy in the world. And he'd show them.

He well knew the weeks of difficult boot training lay ahead of him—weeks of drilling, classroom work, and the hardest kind of physical toughening-up. Then would come *more* weeks of training and hard work before he could hope to be sent to some far-off "Island X" to play his part in the exciting, dangerous tasks that are the everyday lot of men in the Naval Construction Battalions.

With these thoughts, and memories of home, filling his mind, he finally drifted off to sleep.

"G.I."

The next morning, after breakfast in the mess hall, the boys learned that there are many important details that must be attended to when a fellow changes from civilian life to the Navy.

First, back at the Receiving Barracks, the new recruits were lined up in the alphabetical order in which their names appeared on the muster list.

"All right, men," came the order. "Strip down to your shorts."

In front of each boy was a cardboard carton. Into the box he was told to put all his civilian clothes. Then, on a tag which a yeoman handed each recruit, he wrote his name and home address, so that his belongings could be mailed back.

"From now on, everything is G.I.—Government Issue," one of the chief petty officers explained. "Even the haircut you are about to get is G.I.," he continued, with a smile.

Bill Scott glanced down at the line to the "R's" and grinned at Tex Rogers, who had a thick yellow curl that had a habit of flopping down over his left eye.

"Too bad, Curly," Bill said in a loud, hoarse whisper.

"I don't mind the G.I. haircut," Tex whispered back, just as loudly, "but the floor feels like a cake of ice to my bare feet. When do they let us put our shoes on again?"

The chief petty officer glared at the almost naked recruits.

"Quiet, please!" he barked. "I'll do the talking here."

He then explained that each boy would be given a complete physical checkup—heart, teeth, eyes, chest and all the rest—as well as "shots" for yellow fever, smallpox, and tetanus.

"What's tetanus, sir?" someone asked.

"I hope you never find out firsthand." the officer replied.

"It's lockjaw."

One by one the boys fled through the various medical booths for their tests. After that was finished, each was given an empty mattress cover which made a sort of bag about six feet long.

Bill took his mattress-cover sack and, with his checklist in his hand, began the job of drawing his G. I. issue. The list seemed endless. There were Navy trousers, socks, shoes, underwear, caps, jerseys, jumpers, handkerchiefs, leggings, dungarees, gloves, shirts, and so on. As he was given each article he checked it off his list and stuffed it into his sack, until finally the mattress cover was crammed full.

"I feel like Santa Claus at a Sunday-school Christmas party." Bill said to the sailor next to him as they lugged their bulging sacks to the next stop—the fitting room.

Here each boy tried on every piece of G.I. clothing, to make sure that it fitted. Most of Bill's things fitted perfectly, but his shoes were much to small. He had to exchange them for others, a size larger.

The next stop was the stencil room, where each Seabee's name was lettered on his clothing. Such things as Bill's towels, undershirts, and blankets were stenciled "Scott, W" right where it could easily be seen. On other articles, such as trousers, jerseys, jumpers, and caps, a white strip of cloth bearing his name was sewed out of sight on the inside.

When this chore was finished, the recruits were lined up, thirty at a time, for their G.I. haircuts. When it was over, even Tex Rogers' stubborn curl was nothing but a little clump of hair an inch and a half long, strictly G.I.

After the haircut the next stopping place was the photographic room, where each boy had his picture taken for his official Navy I.D. card. Then the boys received their "dog tags," on which their names and serial numbers were stamped. These tags, they were told, must be worn around the neck at all times.

The last, and most important, stop was the Classification Section. Here each incoming Seabee was interviewed by an officer, who examined the boy's recruiting papers and talked to him in order to find out what special skills he might have.

"This is where I get the works." Bill said to himself as he sat down across the desk, facing the officer.

But he soon found that his fears were unjustified. The Classifying Officer was very friendly and quickly gained Bill's confidence. After a few questions about his family and education the officer's eye caught the rating which the recruiting officer had placed opposite Bill's name when he joined the Navy.

"Hmm. Machinist's Mate 3rd Class," the Classifying Officer said, looking up at Bill. "How much do you know about machinery, Scott? Ever run a tractor or a bulldozer?"

"I've never run a bulldozer, sir" Bill replied. "But for the last two summers I've run a tractor on a farm. In fact, I even *built* one."

He told the chief petty officer about the Red Bug (tractor he had built), and how he had sold it just before leaving Danville.

The officer seemed very interested. He asked Bill a number of questions about where he had found the secondhand parts, and how he had cut the chassis and welded it together again.

"That sounds pretty good, Scott," he said when Bill had finished his story. "Our job here is to see that round pegs get into round holes, not square holes. We try to make the most of a man's past experience and fit him in where he can do the most good as a Seabee."

He explained that more than sixty trades and skills were in demand in the Construction Battalions. In addition to operators of bulldozers, power shovels, cranes, and all sorts of heavy equipment, they needed carpenters, sheet-metal workers, plumbers, concrete workers welders, electricians, and so on.

The officer told Bill that the regular Navy ratings which a man might be given did not always describe the sort of work he was supposed to do. For instance, he said, a concrete worker is given a Navy Carpenter's rating. Ships have never carried concrete workers, and so the Navy has no special rating for them. Many Seabees who have a Shipfitter's rating are not actually shipfitters at all, but may be draftsmen, steelworkers, or plumbers. Among the men with the rating of Boatswain there are riggers, divers, and dredgemen. Seabees who operate bulldozers and power shovels carry the rating of Machinist's Mate—a far cry from the machinists who are ordinarily found aboard our Navy ships.

"Because of your mechanical experience I think I'll let that Machinist's Mate 3rd Class rating ride along opposite your name." the officer continued. "But, as you start your training, it will be up to you to prove that you deserve that rating. If you do, you may end up operating a tractor or a bulldozer. Is that what you want to do?"

"It certainly is, sir." Bill beamed.

The officer told Bill he hoped it would work out that way.

"And good luck." he said as he motioned for the next Seabee to come to his desk.

Bill hurried back to Barracks 28 to stow away his gear. As he hung his uniform neatly in his locker, he ran his fingers gently over the two small white initials, "CB," midway between the elbow and wrist of the left sleeve.

He had made the grade. He was ready, now, for whatever came along.

JUDO JUNGLE

In the days that followed, the green recruits discovered that life in the Navy is quite different from civilian life.

Even the time of day is reckoned according to Navy custom, in twenty-four-hour units. Six A.M.—when "Reveille" sounded and the boys tumbled out of their bunks to "hit the deck"—appeared on their written schedules as *0600.* At *0800,* after morning mess, they were mustered. Some of the new-comers were confused, at first, by the way time was reckoned from noon on. One o'clock in the afternoon was *1300* . . . one-thirty was *1330* . . . two o'clock was *1400* . . . three o'clock was *1500* . . . and so on right up to *2359,* which was really 11:59 P.M. by their watches. Midnight was *0000,* and then the *next* twenty-four-hour day began.

Each morning, after muster, the boys were marched off to the huge parade grounds for drill and the manual of arms. They were taught how to carry their rifles correctly, and to execute such orders as "Right shoulder ARMS!" in four quick movements . . . ""Present ARMS!" . . . "Parade REST!" . . . and ""Fix BAYONETS!"

They learned how to form squads at the command "Fall in!"—and to execute close-order drills. In addition to their work on the parade grounds, there were many classroom lectures and training movies. Most of the talks and films, during the first week or two, were about the traditions of the Navy and military courtesy.

It was hard, at first, for Bill Scott and his shipmates to sit in a Quonset-hut classroom listening to lectures when they knew there were so many other things that were more exciting to do.

Usually, while the instructor was talking to them about the importance of showing respect to senior officers, or the strict discipline which the Navy demanded of them, the boys could hear the faint rumble of the diesel bulldozers and power shovels working over at the Proving Grounds, or the sharp crack of the rifles on the firing range. But they knew that a good Seabee *had* to learn the things they were being taught, before he could take his place in one of the regular training battalions.

Not all their time was spent in the classroom or on the parade grounds, by any means. There was a new indoor swimming pool so huge that fifteen hundred men could use it at the same time. Here, in small groups, the recruits were taught to swim, dive, and rescue shipmates "in distress." They learned how to swim fully clothed, even carrying their rifles without getting them wet.

The grand climax came when oil was poured on the water and lighted. As the flames spread and licked upward from the surface of the water, each man had to dive into the oil patch and swim under it to safety.

Tex Rogers was the first to complete his burning-oil test. Then Bill Scott dived in. He broke through to clear water and struck out for the edge of the pool. He pulled himself up beside Tex, who was still puffing and blowing out water like a porpoise.

"Give me the ol' swimmin' hole down home on the ranch, any day." Tex exclaimed, shaking his head. "This burning oil isn't my idea of fun."

Bill laughed. "You've got something there, Tex." he said.

"But what we're learning here today may save our lives if we ever have to abandon a burning ship. I always thought a guy was a dead duck if he had to dive into burning oil. But now I know better."

Gradually the drilling, setting-up exercises and swimming began to toughen their muscles. Boys who had come to the training base soft and flabby could now show swelling biceps, stomach muscles hard as a rock, and a grip like a steel vise.

The toughest part of their physical training was the course in Judo. On rainy days Judo was taught in the gymnasium; but when the weather permitted, classes went out to Judo Jungle—which was a clearing in the woods, beyond the barracks.

The instructor was a barrel-chested sailor known as "Spike" Saunders, who had formerly been the U.S. Fleet's heavy weight boxing champion.

"I'd hate to tangle with that Tarzan." Bill whispered to Slim and Tex as they gathered around Spike in Judo Jungle for their first lesson.

"Me too." said Slim, with a fearsome glance at the instructor's bulging muscles. "He could snap my arms in two like the stem of a clay pipe."

"All right, men," the instructor began. "You're here to learn the tactics of dirty hand-to-hand combat—and I mean *dirty.*"

He paused and pointed to a sign on the tree by which he was standing. The sign read: "Forget the Marquis of Queensbury rules—get in there and FIGHT."

Then Spike went on to explain that Judo was a combination of all that is underhanded in wrestling, boxing, and jujitsu—real jungle warfare at its worst. When he asked for a volunteer to come forward, Tex gave Bill a nudge that threw him slightly off balance. Bill, to prevent falling, involuntarily put one foot forward. The instructor saw that he was the only man who had made a motion to come forward. Bill had "volunteered," against his will.

"You'll do," Spike said, pointing his finger at Bill. "What's your name?"

Bill told him.

"All right, Scott," the instructor said. "First I'll demonstrate the basic principles of hand-to-hand combat."

He grabbed Bill's arm and, almost quicker than the eye could follow the motion, jerked the surprised Seabee over his hip. *Thump.* The next thing Bill knew he was sprawled flat on his back, looking up into the blue sky. He struggled to his feet for another try. This time the instructor told him how to guard against the simple hip-throw.

After several more demonstrations, in which Bill always came out second best, Spike called for another Seabee to come forward. Nick Torrio volunteered. Although he was somewhat more heavily built than Bill Scott, Nick fared no better.

But as the lessons progressed, the boys learned how to do all of the tricks, and the instructor found it harder and harder to thump them around. He patiently showed them how to apply pressure where it would make an adversary helpless. They practiced strangle holds, "backbreakers," eye gouging, hammerlocks, and various methods of striking their opponents with their fists, elbows, knees, and feet, with special emphasis on kicking.

Spike told the Seabees that lessons to follow would include defense against an enemy armed with a knife. He said he would show them how to avoid a slash and to disarm the foe.

"He plays rough." Bill remarked ruefully, as the class disbanded and trudged stiffly back to the barracks.

"So do the Japanese." was Slim's comment.

"You said it." Tex agreed. "I'd rather get my Judo lessons from a U.S. gob than from a buck-toothed Nip on some South Pacific island."

After the Judo lessons the boys actually began to look forward to a chance to rest their aching muscles in classroom work. Learning how to tie all the different knots, how to splice ropes, and how to wigwag signals with semaphore flags was a pleasant relief after a body-thumping session with the Judo instructor.

"HIT THE BEACHHEAD"

Bill Scott had always been fond of guns, even before he joined the Navy.

Back home he had several of them, neatly racked in a cabinet he had built himself. There was a Remington .22, which he used mostly for squirrel hunting . . . a 12-gauge Parker shotgun, for upland game and duck shooting . . . and a 30-30 Winchester, with which he had brought down his first deer when he was only sixteen years old. When he wasn't tinkering with gasoline engines he was usually busy cleaning his guns, whether they needed it or not.

Tex Rogers was a good marksman, too. He told Bill that he could knock off a gopher with his .45 at one hundred yards from his cow pony, at a full gallop.

Both boys were glad when their Seabee gunnery training started. Much of the primary instruction was already familiar to them—how to use the peep sight and open sight, how to time the trigger squeeze with breathing, and how to draw a bead on a target. When it came to taking their rifles apart and putting them together again, Bill and Tex did the job in record time.

Finally they were allowed to work with the new Navy 30-caliber carbines, firing on the 100-yard and 300-yard range from all positions: prone, kneeling, sitting and standing.

Then they received instruction on the deadly .45 [caliber] tommy gun, known officially as the Thompson submachine gun, and the Browning .30 [caliber] machine gun. On all types of firearms the boys not only had many hours of target practice on the range but were thoroughly drilled in the construction of the gun, disassembly, and assembly. Sometimes the gunnery instructor would purposely jam the firing mechanism or do something else to the gun that would cause it to misfire. It was then up to the student to find out what was wrong and correct it.

Some of the class also trained on the famous 20-millimeter Oerlikon antiaircraft gun. For this instruction they went, by bus, down to Price's Neck and fired at a sleeve target which was towed through the air by a Navy plane from the Quonset base. The Seabees who joined the Antiaircraft Section also had to spend many hours in the classroom on aircraft recognition. As pictures of all the different planes were flashed on the screen, they had to learn how to identify them in an instant. They became expert in spotting not only all United States and Allied planes but also German and Japanese planes. In order to pass the course they had to be able to identify each plane in one-twenty-fifth of a second as it flicked across the screen.

There was one type of weapon at Camp Endicott that was entirely new to both Bill and Tex. This was the mortar—a wicked contraption used for lobbing high-explosive shells into the enemy's pillboxes and dugouts. The mortars looked like stovepipes set on bases. When the projectile was dropped down the "stovepipe," it hit a firing pin at the bottom and shot out with terrific force—like a rocket.

One morning, during the second week, the Seabees were told to report for special training in commando tactics. The instructor explained that the first problem would be that of landing on a beachhead from a ship, such as an LST (Landing Ship, Tanks).

The boys gathered about the chief petty officer as he went into the details of the operation. Bill was puzzled. So were the others. The problem involved landing from a ship, plunging into the surf, and all that sort of thing—but they were all standing on dry land. There wasn't a ship in sight—nor a single wave.

But the riddle was soon solved. The officer pointed to a near-by tower about twenty feet high. There were steps leading up to the top of the tower.

"That's the ship on which you are approaching a Japanese-held island," he said. "The problem is to climb down the starboard side, on the cargo net—with rifle and full equipment—into the landing craft, and then storm the beachhead."

He told them exactly how to carry their equipment so that it would not become fouled in the meshes of the cargo net as they scrambled down it.

The Seabees climbed up the steps and stood on the platform at the top of the tower, packed closely together. When the instructor gave the command, they swarmed over the side and down the net like a colony of frightened spiders. At the bottom of the net there was a "landing craft"—a big boxlike affair mounted on springs that made it rock back and forth like a boat on the waves.

The first men to hit the deck of the landing craft dropped the ramp so that all the Seabees could run down it onto the beach.

Each man had a number, which indicated his position in the platoon formation. Bill was number 14. Tex was number 18.

"Remember," the officer called out, "you're supposed to be landing under enemy fire. Deploy on the beachhead and don't break formation."

Shouting and yelling, with bayonets fixed, the Seabees stormed the beachhead. Far up the beach was a wire fence on which signs had been fastened. On each sign was a number, corresponding to the numbers the Seabees had been given. But, in order to reach his station, each man first had to squirm and wriggle his way trough a messy barbed-wire entanglement which the "enemy" had erected to protect the beachhead.

Bill scratched his hands and tore his dungarees in a couple of places as he wormed his way through the tangle of barbed wire. He was certain that he would be among the last to reach his post. But when he finally got there he looked back and found that some of the others were scrambling through the wire. One Seabee was so hopelessly snarled up that two of his shipmates had to dash back through the "enemy fire" to rescue him.

The instructor was not satisfied with the way the landing problem had been carried out, and he let the boys know it in no uncertain terms.

"That was sloppy work." he snapped. "I want you to hit that beachhead as if your lives depended on it—not like a bunch of campfire girls on an outing."

"Perhaps," the chief petty officer continued in a sarcastic tone of voice," you'd rather have me lay down a gangplank for you, with a red plush carpet. I'm so sorry. The Navy is fresh out of gangplanks and red carpet today. Now then," he said briskly, "up you go into that tower on the double quick and do it all over again."

THE COMMANDO COURSE

There was still half an hour before "Lights Out." Bill Scott was sitting on the edge of his bunk, writing a letter home.

"Dear Mom: I'm a movie star now. But I don't think my acting will win the Oscar for the best performance of the year."

"It happened this way. Up here at the training base they have a commando course. It's supposed to be one of the toughest courses in the country. Well, this morning the C.P.O. told us that one of the movie companies was sending a camera crew out to make a newsreel of us going over the course."

"If you happen to see the film one of these days when you're at the movies, watch for *yours truly*. I'm the guy who falls in, when we're swinging over the water hazard on a rope."

And he went on to describe some of the other things that happened that morning when they had their workout on the commando course.

The course itself was almost a mile long, laid over rough terrain and with thirty of the most difficult obstacles that could possibly be built.

"Somebody must have sat up night designing this new form of torture for us." Slim Brown exclaimed as he and Tex and Bill and the other men from Barracks 28 stood in a group, waiting to start.

The newsreel cameraman was already clicking away, running off a few feet of film while the Seabees were getting ready to hurdle the first obstacle.

"All I can say is I'm glad these aren't *sound* movies," Tex remarked. "They'd get nothing but grunts and groans."

One of the chief petty officers motioned for the boys to huddle around him while he explained what they were to do. He described the different obstacles and told them that there would be officers spotted along the course to check them off and grade them.

"Ten men at a time will take the course," he continued, "at thirty-second intervals."

He divided the Seabees into groups of ten. Tex, Slim and Bill were all in the first group. They climbed up onto the platform which marked the beginning of the course.

"All right, Group 1," said the C.P.O., "Go!"

The ten Seabees ran to the edge of the platform and jumped off into space. They hit the ground with a thud, sprawling in all directions. Ahead of them was Obstacle 1, a log wall. Up they scrambled, then down the other side. Now their progress was blocked by Obstacle 2, a barbed-wire entanglement. While a newsreel cameraman clicked away, the boys fought their way under, over, and through the strands of prickly wire.

As they scrambled to their feet again, Tex said, "I know now why we couldn't get enough wire to fence our cattle range last year. They sent it all to Camp Endicott."

Bill glanced back over his shoulder. The second group of Seabees was already under way, shouting and yelling like Indians as they jumped off the platform.

Now came Obstacle 3, a maze of saplings—some of them upright, some crisscrossed, and others laid horizontally. Struggling and puffing, the Seabees worked their way through the jungle of timber—only to dash down a slope to the next obstacle.

The trick, this time, was to get to the opposite bank of a gully across which two slender wire cables were strung, one about six feet below the other. Here, too, a cameraman was lying in wait for victims.

"All we need now is a pair of silk tights and a balancing pole, like a slack-wire performer in the circus," Tex remarked as he waited for is turn to go across.

"And a fireman's net, underneath, to catch us." Bill added.

But they got across without mishap. And on they went, up over twelve-foot stone walls . . . across deep pits on narrow logs . . . through round holes and square holes cut in walls . . . up and over high "chicken-coop" hurdles . . . up ladders and down cargo nets . . . and across rickety jungle bridges.

Where Bill really came to grief was on Obstacle 13, swinging on a Tarzan rope from one bank of a wide water-ditch to the other. Tex reached the ditch first, grabbed the rope, and made it to the other bank safely. As the rope swung back, Bill grabbed it, gave a mighty lunge, and swung out over the water. But suddenly his hold on the rope slipped and—*plop*—into the water he went. To make matters worse, one of the newsreel photographers had selected this obstacle and thus caught the unfortunate Bill on his film.

Bill struggled out of the water and pulled himself up onto the bank, dripping and sputtering.

"And I always thought thirteen was my lucky number." he exclaimed, grinning sheepishly at the cameraman.

"That's O.K., sailor," the newsreel photographer said, with a smile. "It made a swell shot."

"NOT A BOOT IN A CARLOAD"

"What are you looking so happy about?" Bill asked as he peered over Slim Brown's shoulder to see what he was reading. "That's just a schedule for this week's military training, isn't it? Hand grenades . . . bayonet practice . . . another workout on the commando course . . . and all that sort of thing."

"Yeah," Tex chimed in. "What's so good about that?"

"Don't you get it?" asked Slim, grinning like a Cheshire cat. "This is our last week of boot training. And that means two things. *Thing Number One:* next week we start training for the special jobs we came here to learn. Bill can climb aboard his bulldozer and start rattling his kidneys to pieces . . . I can get to be a camouflage expert . . . and Tex can start to learn his Seabee job as diver and pontoon worker."

"You're right." Bill agreed. "But what's *Thing Number Two?*"

"Oh that," grinned his shipmate. "*Thing Number Two* comes along when the Camp is secured, Saturday afternoon. We get our first week-end liberty."

"Yippee!" Tex exploded. "I'd forgotten all about that. Let's go to Providence and celebrate, huh? Just the three of us."

"That's a deal." the others agreed.

And they hurried out of the barracks for the day's first training period, over near the commando course.

This final week of primary military training turned out to be a whirlwind of activity. There were so many things to learn.

They spent many hours learning how to throw hand grenades. They had already had classroom lectures on the construction of grenades, and had seen training movies on their use.

But throwing hand grenades at targets was more fun than studying them in the classroom. Most of the grenades were dummies, although the instructor tossed a couple of real ones at an old wooden shack to show the boys what a destructive weapon it was. When the grenade hit the shack—*wham!*—there was nothing left but kindling wood and splinters.

After a little practice in the field the boys became expert at heaving the "pineapples." The targets were spotted at various distances. Sometimes the Seabees threw the hand grenades from trenches or fox holes, or from behind barricades. And sometimes they threw them on the run, falling prone on their faces to avoid flying fragments. Each man was graded on accuracy, speed and form.

Several times, during the final week of boot training, the Seabees were put through their paces again on the commando course. But Bill didn't fall off the Tarzan rope into the water hazard again, as he did the first time for the benefit of the newsreel cameraman.

In addition to their classes in Judo the men were given instruction in bayonet fighting.

"Most of you will be armed with carbines when you go overseas," the instructor informed them. "But a knowledge of bayonet fighting will come in mighty handy. If the enemy storms your position, he will try to force you out by hand-to-hand fighting. That's when you'll be glad you learned how to use a bayonet."

And he taught them the new slashing style of bayonet fighting, developed by the Marines.

There were also classes in which the Seabees were taught how to detect booby traps and land mines. They were cautioned to watch their step when entering any building from which the enemy had been driven. It was safer to crawl in through a window instead of opening a door, they were told. A favorite trick of the enemy's was to rig up a booby trap so that it would explode as the door was opened. Another enemy trick, especially with land mines, was to bury two mines together, one over the other.

The Seabees also received instruction in first aid. They were shown how to treat wounds and shock, burns, fractures, sunstroke, and frostbite. They learned how to carry injured men, the best method of artificial respiration, and so forth. During the course they were given practical problems to work out like this:

"While building a dock on 'Island X,'" the instructor said, "six Seabees were working near a large pile of lumber. A crane, operating near by, swung over the pile of lumber and happened to bump the top boards. Several pieces of the heavy lumber fell in the midst of the working Seabees, injuring three of them." He picked out three of the students to act the part of the injured men. "One has severe bleeding from the right shoulder, and his left arm hangs limp. Another has a large bruise on his head and is bleeding from one ear. The third has a badly broken thigh, with severe bleeding. Now then, just what first-aid measures would you give the injured men?"

And the boys showed him exactly what they would do.

Another course to which the Seabees paid very close attention was that on the Prevention of Disease and Survival on Land and Water.

"The purpose of this course," said the chief petty officer, "is to help you fight enemies that are just as dangerous as bullets—malaria mosquitoes, ticks, and vermin—and to teach you how to keep alive under very difficult circumstances. After your job is done overseas, we want you to return healthy and ready to take your place again as useful citizens in civilian life."

The boys were shown films on mosquito control, how to purify water for drinking, how to live on a life raft for days on end, how to fight off sharks, how to get along with natives on jungle isles, how to build crude shelters, and how to live on the fruit and other strange foods they might find there.

At the end of that course Bill Scott remarked, "I feel like Robinson Crusoe. I hope that if I ever get stranded on a desert isle I'll have a man Friday to do the work for me."

"Well, don't look at me." was Slim's retort. "Maybe we can get Tex to do the heavy work and pay off in coconuts."

"Sorry," Tex said. "I'll be too busy digging for buried treasure."

As the days went by, the Seabees learned many other things that would be useful to them on "Island X." Over on the practice grounds, near the commando course, they were shown how to erect barbed-wire entanglements and how to dig fox holes, trenches, and other field fortifications "under fire"—*and fast.* They were given that famous jungle knife—the machete—and were taught how to use the deadly blade for gathering brush for camouflage purposes, and to slash a trail through densely wooded terrain.

Finally, using nothing but their machetes, they had to build a footbridge over a twenty-foot ravine. They were allowed fifteen minutes for the job. The bridge was completed in just fourteen minutes, and the entire class walked across to the other side.

On some days, as they dragged their weary bodies back to the barracks, it seemed as though Saturday would never come. But, of course, it did. And when the Training Station was "secured," at 1700, there was a mad rush to get their blue uniforms out of the lockers. Bill, Tex, and Slim got into their shore-leave gear "on the double," and shined their shoes until they glistened. Then, with their caps placed at just the right angle, they nipped out of the barracks to join the other Seabees who were bound for Providence.

They showed their I.D. cards and liberty passes to the guard at the main gate and then ran pell-mell over to the Navy busses, which were already loading up.

When a sailor gets his first week-end liberty, there's not a minute to be wasted—and the bus drivers knew it. They clanked into gear, swung out onto the northbound highway, and off they went.

Bill Scott settled back in his seat.

"There's not a Boot in a carload," was the thought that went through his mind. "We're all full-fledged Seabees now—even me."

And he laid his left arm on the seat rest so that anybody who wanted to could plainly see those two white initials—"CB"—on his sleeve.

DIAMOND AND RING

In training its fighting men the Navy has always believed that "All work and no play makes Jack a dull gob." And that goes for the Seabees, too.

Wedged in between their training schedule and the nightly bugle call, "Lights Out," there was many an hour of activity that was quite different from the routine training for life on "Island X."

Tex Rogers and Bill Scott both played instruments in the Seabee swing band. Tex could make a guitar almost sing, while Bill's specialty was his "old slide horn." At some of the Training Station dances, when Bill got "hot" on his trombone, everyone would stop dancing and just stand there on the floor listening. But most of the time he used the "sweet" technique made famous by his idol, Tommy Dorsey.

Camp Endicott had a baseball team, too. Already, on their own diamond, the Seabees had played and beaten the Quonset Naval Air Station nine, the Coast Guardsmen from New London, and the Army team from near-by Camp Edwards. The only game they had dropped was the day they took on the boys from the PT Boat Training Base at Melville, across Narragansett Bay. Slim Brown, who played first baseman for the Seabees, claimed that he lost that game personally. The score had been tied in the first half of the ninth inning, with two men on bases, when he came up to bat—and fanned out. In the second half the PT boys got a homer.

There was boxing, too—with hair-raising slug fests in the gym whenever the Seabee boxing team went into action. Most of the meets were with boxing teams from other Naval Stations whose training, like that of the Seabees, included Judo. And when Judo experts "mix it up" in the ring, that's a sight worth watching.

In the first session, with the submarine crews, down in New London, Bill Scott slugged it out with a heavyweight from Brooklyn and got knocked cold in the first minute of the second round. Since then, however, he had won all his bouts—two by a knockout and three by a decision

Even now, whenever any of his shipmates kidded him about that Brooklyn roundhouse swing he had caught on the button, Bill would shake his head sadly and say, "That was the quickest 'Lights Out' that ever came my way."

Sometimes, when the boys were caught up with their studies and had a free evening, they would wander over to the recreation hall to "see what's cooking." There they could listen to the radio or play the latest records. Or, if they were feeling real energetic, they could settle down to a fast and furious game of dominoes or checkers.

The dart game was a great favorite with many Seabees. It was called "poker" darts because the scoreboard was made up of playing cards. The object of the game was to try to hit the cards that would make the best hand.

The three shipmates from Barracks 28 could usually be found at the bowling alleys whenever they had any time to spare. They all bowled about 150 and were pretty evenly matched. But one evening Slim put the place into an uproar by bowling a perfect 300 game. The public relations officer heard about it and sent the Station photographer over to take Slim's picture, poised with a bowling ball in his hand. The picture appeared in the sports section of the Providence newspaper.

Slim bought a dozen copies of the edition and mailed them home to his family and friends.

"What a fluke." he wrote his Dad. "I never bowled over 180 before in my life."

There was a library at Camp Endicott, too. Bill often dropped in to draw out a book.

"It's funny," Tex commented one evening, "how often Bill goes over to the library these days. This sudden thirst for culture wouldn't have anything to do with the fact that the librarian is a pretty Wave, would it, Bill?"

"Pipe down, sailor." Bill retorted, blushing in spite of himself. "For a gob who can't read anything deeper than the comic strips in the newspaper, it seems to me *you're* quite a regular visitor at the library, too."

There were several hundred Waves at the Seabee Training Station. They lived in special barracks of their own and were nicknamed "Honey Bees." They did very important work aboard the Station, keeping records and doing a great many other necessary routine jobs. This was work that otherwise would have been done by sailors, who were more valuable to the Navy as builders and fighting men.

On Saturday nights there was always a good movie in the camp's auditorium, and sometimes radio shows were broadcast right from Camp Endicott. The boys got a big thrill the time smiling, sweet-voiced Kate Smith broadcast from their auditorium. When she was making a tour of the Station before her program went on the air, she even climbed aboard a bulldozer and let them take her picture.

A NEW BATTALION IS BORN

The Seabees' basic training was over. And with the end of their boot training came moving day.

The boys in Barracks 28 gathered together their uniforms and other gear and carried all their belongings over to Barracks 54, in Area B. They had been assigned to the newly formed 125th Battalion, which was to be quartered here.

"A much nicer neighborhood." was Bill Scott's comment as the Seabees made themselves at home in their new barracks.

"I can't see any difference at all," Tex said, surveying the rows of bunks that lined the wall. "Same old view—the bunks are just as hard—even the same old neighbors came along."

Somebody heaved a duffel bag at the Texan. It hit him squarely and sent him sprawling off his bunk onto the floor.

"What are *you* beefing about?" Slim Brown asked. "We thought that when we moved we'd get rid of you and your guitar. But no such luck."

As a matter of fact, although none of them would admit it, the boys had been afraid that they might have to split up when they shifted to the new Battalion Barracks. They were really pleased to find that once again they would be bunking in the same quarters again.

The 125th Battalion, like all those which had gone before it, consisted of about one thousand men and thirty-two commissioned officers. From then on, although their toughening-up routine would continue, the Seabees would train more as a unit, in companies and platoons and under the same officers who would command them when they went overseas. Starting now, elected men in the Battalion would be sent to special schools to learn they jobs they would do on "Island X." They were each issued technical-training manuals, with schedules for the weeks ahead.

There were about forty-five different schools on the Station, each classroom given over to a special subject in charge of a chief petty officer.

"They weren't kidding when they told us that Camp Endicott was the most complete vocational school in the United States." Bill exclaimed as he glanced through the technical-training manual, that first day in Barracks 54.

"You said it." Tex agreed, and he ran his eye down the long alphabetical list of subjects that were taught at the Training Station.

There were classes on air compressors • arc welding • blasting • bridge building • bulldozer operation • camouflage • crane operation • dam construction • deep-sea diving • Diesel engines • distilling and purifying water • dock building • electricity • fire fighting • gasoline engines • generators • grading roads and airstrips • ice machines • ignition systems • lubrication of equipment • marine engines • pile driving • pontoons • power-shovel operation • pumps • radio • refrigeration • road building • soil testing • tractor operation • transformers • vulcanizing • and many other subjects.

When the new Battalion assembled in the auditorium that first evening for a talk by their Commanding Officer, the boys learned that the most important subject of all wasn't even listed on their schedules. That subject was Seabee "savvy."

"You must learn to be resourceful," the C.O. said "Many times, when you are working on a job at some far-off base, your equipment may break down and you won't have the spare parts you need to put it back in running order. This doesn't mean that you stop work. It means you *make* the new part you need. Out of what? I don't know. That's where Seabee 'savvy' comes into the picture. And if you haven't got certain supplies you need, you just don't sit back and wait for them to come. You find something else that will do the job just as well."

He told them about one battalion that had landed on a remote South Pacific island. The Seabees were busy stringing up a telephone line and power line, but discovered that they had no insulators for the wires. Suddenly one of the men spotted some boxes containing empty Coca-Cola bottles. With a little tin, taken from the cans their food had come in, they rigged up the glass bottles as insulators and they worked perfectly."

"On another island," the Commanding Officer continued, "the men were building shelters and huts to live in when they found that the roofing material had not arrived. So, taking some empty gasoline drums, they sliced them in half, flattened them slightly, and overlapped them to make a sort of Spanish-tile roof that was absolutely watertight."

He told the boys that Seabee Battalions are rough-and-ready outfits, able to find ways to do the impossible.

"I have no use for red tape," he said. "There's always a short cut, and that's the way a Seabee gets things done."

He concluded his talk by telling them about a crew of Seabee builders who were blasting a roadway out of volcanic rock at a Pacific base. The work was going very slowly because they had to drill eight-foot holes into the rock with a rock drill—holes in which to place their charge of dynamite.

"When a job goes slowly, Seabees get plenty mad." the C.O. went on. "And these boys were mad as hornets. But suddenly, along came a Marine in his General Sherman tank. The Seabees took one look at the cannon on the tank. It gave them an idea. They asked the tank gunner if he would pump a few armour-piercing shells into the rock cliff. He obtained permission from a Marine captain, and did so. Each shell went about ten feet into the rock and made a perfectly drilled hole for the dynamite. In no time at all the entire cliff was blasted out of the way, and the Seabees pushed onward. Because they used their heads, Army Jeeps and supply trucks were moving inland on a new road a whole day sooner."

The Commanding Officer's talk put new meaning into the Seabees' famous slogan: *"Can do, will do—did!"*

THE RED BUG'S GRANDPAPPY

Bill Scott hurried down the main street of the Training Station to a big building on which was a sign that read: "Heavy Equipment Building."

It was time for his first class in technical training. But, when he entered the building, Bill saw the strangest classroom he had ever been in. There was no desk for the instructor, and there were no seats for the students. Instead, there were long workbenches on which were pistons, valves, crankshafts, bearings, and all the other parts that make up a Diesel engine. On a test stand was a whopping-big Diesel engine completely assembled and hooked up to run, just as it would ordinarily be installed in a bulldozer.

This was the kind of classroom in which Bill felt right at home. While the other Seabees came straggling in, he wandered over to the corner to look at a giant contraption which stood there. It was a bulldozer—a large Caterpillar D-8—and it was by far the biggest tractor Bill had ever seen. He touched the heavy caterpillar track, with its deep-cleated shoe plates. This was what gave the bulldozer its terrific traction.

Bill smiled as he thought how funny his homemade Red Bug tractor would look standing alongside this big Caterpillar.

"Golly." he exclaimed. "This D-8 is the biggest darn hunk of machinery I've ever seen. It sure is the Red Bug's [Bill's home-built tractor] grandpappy—the grandpappy of all the tractors that were ever built."

What a thrill it would be when he could climb aboard that bulldozer and operate it.

Just then the instructor came in. He was a tall, powerfully built chief petty officer. He greeted the Seabees and walked over to the Diesel engine, which stood in the center of the room. The boys formed a group about him.

The classroom itself was really just a walled-off section of floor space at one end of the Heavy Equipment Building. And what strange noises were coming from the other of that ten-foot partition. Hammers clanging on steel . . . the hiss of welders . . . and, suddenly, the deep-throated rumble of a big Diesel engine.

"The racket on the other side of that bulkhead comes from the repair shop," the instructor explained. "That's where we repair our bulldozers and other heavy equipment. Later on you'll all have a chance to put in some hours there. But first," he said, laying his hand on the Diesel engine in front of him, "let's find out what makes this baby tick."

Using large charts and cutaway pictures of the engine, the Chief carefully explained the different parts of the huge power plant. He pointed out that a Diesel engine, unlike an automobile engine, has no spark plugs. The fuel is exploded by the intense heat generated by the compression stroke, or upstroke, of the piston.

In addition to the pictures, he showed them the actual parts, one by one, spread out on the workbenches. The first day's work ended with the showing of several movies on the operations of Diesel engines.

The following day, and for several days more, the Seabees were given actual experience tearing down the classroom Diesel engine and assembling it again. This was the part of the course Bill liked best. It meant that once again he had a wrench in his hand and was finding out the wonderful "inside story" of an engine. And what an engine that Diesel was.

Aided by charts which showed them exactly how to do it, the Seabees took the entire power plant apart—and then they put it together again. They studied the combustion system and lubrication system, and found out how to make the fine adjustments that would keep the engine running at top efficiency.

The Diesel engine was so big that it had *another* engine to start it. The starting engine was a small 30-horsepower gasoline engine called a "pony engine."

It was mounted beside the Diesel engine. You had to crank the pony motor. After it began to chug away, you threw a small lever that geared it up to the Diesel engine. When that happened, the Diesel would give out deep sighs, inside its cylinders, like a sleeping giant. Suddenly—*brrp! brrp! brrp!*—and it opened up with a thundering roar. Then you cut the switch on your pony motor, and that's all there was to it.

Finally, as the course went on, the Seabee mechanics became so expert at tearing down the Diesel engine and assembling it again that they could almost have done it blindfolded. Sometimes the instructor put "bugs" in the engine to give them practice in trouble shooting. He would purposely put some part out of adjustment and the boys would have to find out why the engine wouldn't start, or why it kept stalling. When they found out what the trouble was, they had to correct the condition.

"I know you fellows all want to finish your shopwork and get outside to learn how to operate a bulldozer," the Chief said. "But what you're learning now is very important. A good bulldozer man must know what's going on *inside* his machine in order to be a good operator. You may be glad, someday, that you took the time and trouble to know your engine inside and out."

And he told them about an experience he had been through when his Battalion was overseas.

"It was on Guadalcanal," he said, "I was down at the far end of the air strip with my bulldozer, filling in some bomb craters, when my engine sputtered and died. I figured that I had a fouled injector and I jumped down to dismantle it and see what was wrong. Just then a Japanese sniper, back in the jungle, started taking pot shots at me. He wasn't a very good marksman or I wouldn't be here today telling you about it. But believe me, I didn't waste any time getting that bulldozer engine going again and scuttling out of there. That neighborhood just wasn't healthy—not even for a *fast* mechanic.

When the boys finished their course on Diesel engines they tackled the bulldozer itself. They learned how to force out the kingpin that held the caterpillar treads together, so that they could replace broken plates in the track. They tore down and reassembled the transmission and the huge differential gearbox, and became thoroughly familiar with the master clutch, the steering clutches, and all the different controls. Although they could not actually operate the massive bulldozer on the limited floor space of the classroom, they got to know that eighteen-ton piece of equipment from top to bottom and inside out.

Their final hours in the Heavy Equipment School were spent on the other side of the "bulkhead," actually repairing bulldozers that were in use at the Training Station.

It would not be long, now, before they would go out to the Proving Grounds—those desolate acres of sand dunes, rocks, swamp, hummocks and gullies where they would actually learn how to put a bulldozer through its paces. That was the day Bill Scott was waiting for.

DIVING SCHOOL

While Bill Scott had been attending classes in the Heavy Equipment School, the other Seabees in Barracks 54 had been busy, too.

Slim Brown was spending most of his time learning the secrets of military camouflage and Tex Rogers had been going to the Diving School each day.

The first morning he reported for instruction as a diver, Tex kept his fingers crossed for luck. It was common knowledge on the Station that out of every ten Seabees who tried out for Diving only four could pass the special medical examinations. Where most of them washed out was in the compression chamber. This was a big horizontal tank, like a steam boiler. Each boy was sealed in, and the pressure was gradually increased. Then it was brought back to normal again, and the boy came out. Most of them just couldn't take it.

Tex heaved a sigh of relief when the doctor listened to his heart action through a stethoscope, then thumped him on the chest and said, "O.K., sailor—you'll do."

The diving instructor was a happy-go-lucky Irishman by the name of Hogan. He had been a diver in civilian life for over fifteen years. Every since Pearl Harbor he had been a chief petty officer in the Navy, teaching others to dive.

When Tex and the other Seabees in the new class reported for instruction that first morning, the Chief looked them over with a critical eye and made a little speech.

"Every good diver is proud of his trade," he said, "My father before me was a deep-sea diver, and the best there was. He taught me everything I know. That was a lucky thing for me, because it's a tradition among divers to guard the secrets of their trade. There are lots of things a diver wouldn't tell you—not any more than a good magician would tell you how he does his tricks. But, laddies"—he smiled—"now there's a war on. So old Tom Hogan is going to tell you everything he knows about this business of diving. It breaks my heart to give away *some* of the tricks of the trade I'm going to tell you—but

Uncle Sam's Navy needs good divers, and that's what you'll be when ***I*** get through with you."

For the first few days the Seabees spent most of their time becoming familiar with the equipment they would have to use. The Orco diving mask was to simplest gear, for shallow-water diving. It looked something like a gas mask and fitted tightly around the face. The Miller Dunn diving hood was a more elaborate affair, made of metal with glass windows in it.

The deep-sea diving equipment was the queerest of all, with its thick canvas suit, heavy boots weighted with metal, and the headpiece which rested on the diver's shoulders.

There was a life-size dummy in the classroom, completely dressed in deep-sea-diving equipment. The dummy was nick-named "Oscar." The instructor said that, before he found Oscar, he used to put the diving gear on one of the students in order to demonstrate the different parts and how the air line was attached. But one student got so hot, wearing the heavy gear in the classroom during the lecture, that he fainted. That was when Oscar came to the rescue. He had been serving faithfully now for many months.

The biggest thrill in the diving course came when the Seabees first went down, wearing the Orco masks. There was a huge round practice tank in the Diving Building. It was fifty feet across and twenty-six feet deep, and held 500,000 gallons of water. On the outside of the tank, at eye level, was a glass porthole through which the instructor and the other men in the class could watch the Seabee who was working under water inside the tank.

Tex was a little nervous the first time he climbed down the ladder to the bottom of the tank. But he soon got over the feeling, when he found that he could breathe all right. And he could see very clearly through the windows of his mask. He could even see the Chief peering at him through the peephole. His only worry was to keep the life line from fouling as he walked around.

As the boys became more accustomed to being under water, they were given various problems to do. A thick wooden dock piling had been sunk, upright, at the bottom of the tank. Each boy had to saw off a section of it. A much more difficult problem was to disassemble a puzzle made of fifty-two pieces of iron pipe and put them back together again, under water. That was worse than any jigsaw puzzle Tex had ever worked on, but he finally did it.

In addition to diving, each student also had to take his turn standing "topside" and tending the lines and ropes for the man who was under water.

There were many things to learn before a man could hope to qualify as a Seabee diver. As usual the boys were shown many training films, so that when they had to try out something new for the first time they knew something about it. They found that there was even a knack to recovering their balance from different positions. Walking and bending over against the pressure of the water wasn't the easiest thing in the world to do, especially in a heavy diving suit and in boots that felt like a couple of big lead chunks.

After several weeks in the practice diving tank, the class went down to the harbor and got some real experience working from a regular Seabee diving barge. Here they learned the trick of working and moving about in deep mud, how to use a rock drill under water, and how to place a dynamite charge for under water blasting.

The hardest thing to learn was the proper use of welding and cutting torches under water. It was the strangest sensation Tex had ever experienced, being down about thirty feet in murky water, dressed like a man from Mars, and holding an oxyhydrogen torch which spurted blue flame.

It was exhausting work, too. For the first week or so Tex was too tired after evening mess to join his shipmates when they went over to the recreation hall or to the movies.

Bill kidded him about it.

"For a guy who spends most of his time walking around at the bottom of the harbor, you sure do a beautiful swan dive for that bunk of yours these nights."

But Tex didn't care. He noticed that any Seabee who had qualified for diving, as he had, was considered sort of "special" around the Station.

That made him feel rather proud. But more important still, Tex knew that, when the Battalion finally hit "Island X," there would be some tough assignments for the underwater boys and he would be ready for them.

NOW YOU SEE IT—NOW YOU DON'T

"Hey, Slim, how's the Camouflage course coming along?" Bill asked one evening, soon after they had started their technical training.

"Swell," Slim replied. "But it's a lot harder than I thought it would be."

"Harder?" asked Bill. "All you do is paint things green so that they blend with the grass and trees, and brown so they can't be seen against a desert background, isn't it?"

"That's what *you* think." was Slim's retort. "Here—take a look at this."

He showed Bill the outline of the subjects he was studying in order to become a Seabee camouflage expert. Just the *list* of things he had to learn filled five typewritten pages.

"Gee," Bill exclaimed, "that course is no pipe after all."

It certainly wasn't a "pipe," but it was one of the most interesting courses taught at the Station. The day the Camouflage class met for the first time, the instructor related an amusing incident that took place when he was with his Battalion in the Southwest Pacific

"The Marines had already taken half the island and had moved inland to mop up the Japanese," the chief petty officer told them. "Our Battalion had been working a couple of days and nights cleaning up the wreckage; building living quarters, storage tanks, repair shops for our equipment; and leveling off the bomb-pocked air strip so that the Marine fliers could use it. As fast as we built something we camouflaged it, because Japanese air patrols kept flying over to take pictures of our installation."

The following day another company of Marines came ashore from the landing barges. With them was a famous American artist who had come to the South Pacific at the Navy's request to make a series of pictures showing what a Japanese island looks like after it has been stormed and taken by the Marines. The artist stepped onto the beach and stood there in amazement.

"This isn't the sort of thing I came here to draw," he said in a tone of disappointment. "You Seabees work too fast. In two days you've built a new island base—*and hidden it from sight.* I'm going over t the other side of the island where the Marines are still fighting."

"You'd better hurry, sir," a Seabee lieutenant said. "Some of our boys are over there already, where the fireworks are going on, cleaning up the mess and building new installations."

And that, the instructor told the new class, was the way they'd have to work when *they* got to "Island X."

First they learned the three chief methods of camouflaging—*hiding* a thing by covering it, *blending* it with the surrounding landscape, and *deceiving* the enemy by making it look like something else. They were shown many movies which explained each method, and they listened to lectures on how to use various materials for camouflage.

A good deal of the instruction was given right out in the fields and woods and on the sand dunes of Camp Endicott. The Chief showed the boys how to weave foliage into wire netting to cover gun emplacements, fuel dumps, tanks, planes, and fox holes. They were taught how to use portable paint-spraying equipment to help conceal anything which the enemy should not see.

One part of the course reminded Slim of some of the games he used to play when he was a kid. It was called "Personal Concealment." The Seabees were shown how to daub their faces and uniforms with various paints so that they would blend with their background, whether they were fighting in the tropics, on the sandy desert, or on snow-covered Arctic terrain. There was even a knack to cutting small leafy branches to stick through the web netting that covered their helmets.

When they had learned all the tricks of personal concealment, the class divided into two teams. One team would hide among the rocks and brush, and the other team would try to find them. By the time the boys had become so expert at camouflaging themselves that sometimes they could not be seen even at a distance of only twenty feet.

One morning Slim had concealed himself in a bush beside a big rock. He hunched in his cover, his tommy gun clenched in his hand for instant action against the "enemy." He watched as one of the "enemy," a Seabee by the name of Bob Nash, came closer and closer to his hiding place. Slim didn't move a muscle. Nash came closer and closer, peering this way and that. Still he didn't see Slim. Suddenly he stepped right on Slim's foot. He felt it move, and jumped back as though he had stepped on a rattlesnake.

"Hey!" Slim exclaimed as he straightened up and came out of his hiding place. "You're supposed to *find* me—not trample me to death.

Bob Nash laughed. "You sure had me fooled." he said. "I didn't think there was anyone within fifty feet of me."

There was another way in which camouflage was used to deceive the enemy. The Seabees learned how to build and use dummy planes, tanks, trucks, and even dummy houses.

The instructor told his class how the British, at one point in the North African campaign, used such decoys to fool Rommel.

"They built flimsy dummy tanks, out of wood and canvas," he said. "They even had a wooden cannon sticking out of each gun turret. Then they set the dummy-tank bodies on Jeeps and drove them across the desert, slowly and in formation, so that the Germans could see them coming, through their binoculars. Rommel, observing these hundreds of tanks moving to outflank him shifted *his* tanks to engage them in battle.

"That was just what the Allies were waiting for," the Chief continued. "They immediately rush their *real* tanks around Rommel's other flank and knocked out most of his panzer division from the rear. In the meantime the dummy tanks had done their job. The Jeep drivers shifted into high and raced back to the Allied lines."

Some of the planes and trucks which the Seabees learned to build out of wood and canvas looked so real that they could scarcely be told from the real thing. And from the air any enemy scout would have been completely fooled.

The dummy equipment was usually not placed out in the open, where it could be plainly seen. Each truck and plane would be very *lightly* covered with leafy branches, so that the enemy planes would think it was real equipment which the Seabees actually wanted to protect.

Another part of the task was to make sure that decoy equipment was placed in a spot far removed from the *real* trucks and planes, so that enemy bombers would waste their explosives on the decoys instead of bombing the real thing.

In order to find out how good their camouflage tactics were, the Seabees were shown photographs that had been taken from the air. After studying these photographs, they went out in the field again and practised camouflaging roads and building decoy roads that would fool aerial scout planes. Airstrips and runways were the hardest things to hide from enemy eyes but there were even ways to do that.

At the end of several weeks spent learning how to hide things and how to conceal even *himself,* Slim told his shipmates one night that he was beginning to feel like an "invisible man."

"Well," Tex drawled as he sat on the edge of his bunk unlacing his G.I. boots, "with a face like yours, I don't blame you for wanting to hide it."

Slim made a flying leap across two bunks and landed squarely on Tex. They fell off the bunk and rolled on the floor together. Several other Seabees joined in the friendly fracas. Fortunately, just as it seemed as if Barracks 54 would need a riot squad, the bugle sounded "Lights Out," and in thirty seconds all was peaceful and quiet once more.

KEEP YOUR HEAD DOWN

When the boys of the new Battalion started their technical training in Diesel engines, diving, camouflage and so on, some of them thought that perhaps their training as fighters was finished. But it wasn't.

No Seabee at Camp Endicott ever had a chance to forget that the motto, "We build—we fight," meant exactly what it said.

Several times each week they would pile the huge trailer bus known as the "Swoose" and go out to Sun Valley for advanced combat training under battle conditions.

One of their first problems was to learn how to maneuver in a smoke screen, under fire.

"Sometimes the enemy will lay down smoke in an attempt to confuse you and destroy your morale," the instructor explained. "So it is very important for you to learn how to carry on, even in a smoke screen."

On the assault course at Sun Valley there was a barbed wire maze with openings spaced at irregular intervals. The men were formed into squads and, while the smoke generators laid down a heavy layer of "soup," each squad had to run across the field and work its way through the crooked lanes formed by the barbed wire. The men soon became so accustomed to the smoke screen that they could work out this maneuver in fine style. At last the instructor was satisfied that they would not become panicky if they ran into an enemy smoke screen in actual combat.

At Sun Valley the Seabees were also given training in chemical warfare, and were taught how to protect themselves from the gas attacks by detecting the odors of the various gases.

"If you suddenly smell geraniums," the chief petty officer said, "don't waste any time trying to find where they are. *That* smell means Lewisite gas—so whip out your gas masks immediately and put them on."

He gave them an opportunity, too, to find out that an apple-blossom smell tear gas, and the odor of fresh-cut corn stalks meant the presence of deadly Phosgene gas.

"I always knew that this big nose of mine wasn't meant for beauty," Slim remarked to Bill on the way back to the Station that afternoon. "At last I find that it has a practical use."

The Seabees had long ago passed the point where they needed training as *individual* fighters. Their advanced military training was for the purpose of teaching them to maneuver and fight as *units*—in squads, platoons and companies.

Bill Scott was a squad leader. He was responsible for the way his men carried out each problem that was given to them. The instructor would say, "Scott, your rifle squad has been detailed as an advance guard. Your orders are to establish contact with the enemy." Then he would ask Bill a number of questions: "What commands do you give your squad?" . . . "What communications do you maintain with the elements to your rear?" . . . "What is the formation of your squad?"

Bill also had to know exactly what to do when his scouts would signal back, "Enemy sighted—approaching in deployed formation."

Before carrying out maneuvers such as this he would go into a huddle with his men, like a quarterback on a football team. Then they would do the job they were supposed to do, under the ever-watchful eye of the instructor. Things didn't always work out the way they were supposed to. When that happened, it was up to Bill, as squad leader, to do some fast thinking and get his men out of trouble.

Problems like these were easy, but some of the advanced training was really rugged—especially the sham battles. In order to test the morale of the Battalion under fire, the boys were formed into squads and had to work their way over the assault course as skirmishers—under circumstances almost like those in actual battle.

This was a test that took plenty of courage. Land mines had been planted, here and there, all over the assault course. The Seabees knew that the instructors would keep exploding one mine after another, by remote control. And there were booby traps scattered over the course to make trouble for the careless skirmisher. In addition to these hazards a smoke screen would be laid down, and expert gunners in the rear were ordered to lay a ceiling of gunfire over the course, three feet above the ground, using live ammunition.

One of the chief petty officers carefully explained all the hazards of the course to the Seabees, making certain that they understood the dangers of the problem to work out.

When he told them that the terrain would be raked with gunfire, at a height of only three feet, he added, "We aren't fooling, this time. Remember this is no game—it's the real thing. It's as close as you'll come to actual combat operations this side of 'Island X.' So when I say 'KEEP YOUR HEAD DOWN!' *I mean it.* That's the only way you'll be able to come through this maneuver in one piece."

When he gave the signal, the boys flung themselves head-long onto the ground and the fireworks started. The smoke generators poured out a heavy, low-hanging pall of smoke which blanketed the ground like a dense fog. The gunners began to fire over the heads of the boys. Squirming on their bellies, the Seabees started to crawl over the course, inch by inch, with their carbines in their hands.

Bill Scott led his squad in this difficult, nerve-racking maneuver, constantly on the alert for concealed land mines and booby traps. Now the bullets were whining and crashing through the underbrush, a never-ending reminder of the instructor's warning to keep their heads down.

After he had crawled about ten yards through the brush, Bill found the progress blocked by a barbed wire entanglement. He examined the barricade closely for the presence of booby traps.

"Well, here goes." he said said.

Cautiously he pried the two bottom strands of twisted wire apart and wormed his way through. Suddenly there was a terrific explosion, close by. Dirt and stones and twigs came showering down. It was a land mine, set off by an instructor at the rear.

"That was too close for comfort." Bill muttered.

He adjusted his helmet and continued his snakelike progress over the assault course. A quick glance to the rear told him that the other men in his squad were following closely. Just then—*ping!*—and a bullet smacked into a tree trunk, not more than two feet from him.

Now there was a shallow trench directly ahead of Bill. He wondered whether it might not be better to circle around it. But he decided against doing that. Swinging both legs around so that they were parallel with the trench, he rolled into it; then cautiously he clambered up the other side.

The smoke was getting thicker now. Another mine suddenly blew up, about thirty yards ahead of Bill—then another, off to the right. He inched forward a short distance to a big rock and lay there a moment, hugging the ground, to keep his breath. Just then he heard a rustling in the leaves near by. He looked over. It was Tex Rogers, crawling slowly in his direction.

"Hi, Tex." Bill called out above the noise of the gunfire.

Tex heard him and squirmed over toward the rock. As he hunched forward to flop down beside Bill, he carelessly raised his head. There came a metallic *ping!* and his helmet fell off.

"Hey!" Bill shouted. "Keep your head down!"

Tex was flat against the ground now, thoroughly scared. His face was white. He reached out one hand and got his helmet. When he saw the dent where the bullet had nicked it, he a gave a low whistle.

"Golly, that was a close call." he exclaimed.

"It was a lucky thing for you that you had your helmet on." Bill agreed. "Let's get going now. The enemy barricade is only about fifty yards farther on."

Just as they were about to get under way again, the firing stopped. A minute later the bugle sounded, signaling that the infiltration maneuvers were over.

Now the Seabees had learned, for the first time, what it was like to go into combat under fire.

TAMING THE "CAT'

"In civilian life a contractor figures on a year and a half to two years to train a good bulldozer operator," the instructor said. "Here in the Navy we don't have that much time. We figure in days and weeks, instead of years. But, by using all the short cuts we know about and by constant drilling, we'll send you out of here with good basic training as bulldozer men. The experience you'll get overseas will soon do the rest."

The instructor was a big husky chief petty officer who had been a construction man all his life. He had operated a power shovel on the Panama Canal, and a bulldozer on Grand Coulee Dam. He had built roads, dams, canals, and airports from Alaska to Peru. What he didn't know about construction and earth moving just wasn't worth knowing. And now, as boss of the Seabee heavy-equipment Proving Grounds, it was his job to train the boys who would soon be slashing out roads and military airports on jungle islands halfway around the world.

Today he was giving his introductory talk to the Seabees who were to become heavy equipment operators in the newly formed Battalion. Bill Scott and some of his shipmates sat in the classroom in the big building which served as the Proving Grounds School. They listened intently to everything the Chief said, for they knew there was very little time to learn the ropes.

"If you have confidence in your bulldozer, you can make it do anything you want it to," the instructor continued.

He showed them movies of a giant "cat" knocking over trees fifteen inches in diameter, just as if they were wheat straws . . . moving thirty-ton boulders around like pebbles . . . and shoving huge three-yard loads of earth this way and that.

After the classroom talk the Seabees went outside to try out the machines for the first time. And what a line-up of equipment there was. It was the strangest parking lot Bill had ever seen. There were Allis-Chalmers bulldozers, big and little: International bulldozers, all painted red; and all sizes of Caterpillar bulldozers, painted yellow. The machines were lined up in a long row, ready for the day's work.

Stepping over to a medium-sized International bulldozer, the Chief explained that it had a starter instead of a pony motor. He demonstrated how to start it. All you had to do was turn on your gasoline valve, throw the switch, and press the starter button. Then, as soon as the Diesel got going, you cut in your heavy-fuel line. The engine would pause for an instant, and then burst into a terrific roar as the Diesel oil got to the cylinders.

From the International they walked down the line to a big yellow eighteen-ton "cat." Even though Bill had worked on one of these giant machines over in the Heavy Equipment School, he still got a big thrill at the tremendous size of the huge Caterpillar D-8.

"These babies cost about nine thousand dollars each," the instructor said. "They're tough and can take a lot of punishment. But, like any other piece of machinery, they can be wrecked by *abuse.*"

He gave the boys a talk on the safety rules which should always be followed, for the sake of both the operator and the machine. The first thing to do, before starting the engine, was to check the oil level in all compartments.

"Then throw the lever that disengages the master clutch and make sure your gearshift lever is in neutral," the Chief explained. "All right—one of you start this engine."

He singled out a Seabee who happened to be standing near the front of the machine. His name was Carson.

After checking everything he was supposed to check, Carson got around in front of the "cat," between the big pusher blade and the engine, to crank the pony motor. He was a big, powerful fellow, and very eager to show he knew his way around a bulldozer. He gave the crank a hefty tug and spun it completely around.

"Hey! Hold it!" the Chief shouted, motioning for Carson to stop cranking. "That's a perfect demonstration of how to break an arm. Never spin the crank, Carson. If that engine kicked, the crank would snap your arm in two like a twig. I thought you knew better than that. Now try it again."

Two more short tugs on the crank started the pony motor popping away. Carson cut-in the Diesel, throttled down, and stepped back.

"All right," the Chief said as he engaged the master clutch. He gave the boys a few pointers on what to do next. First you had to check the fuel and oil pressure on the gauge.

"If it doesn't come up in about half a minute," he cautioned them, "shut off your engine and find out what's wrong."

He told them never to start working the engine until it had warmed up to about 180 degrees.

"Now then, who's the first to try out?" he asked.

Bill was standing next to him.

"I'd like to, sir," he said.

"O.K. Up you go." the Chief said.

Bill climbed up over the steel track into the driver's seat. The Chief followed and sat down beside him.

What a lot of pedals and levers there were to work. To steer the "cat," you had to pull back one of the steering-clutch levers and press a foot pedal. This made *one* track move. To move the other track, you had to do the same thing on the *other* side. And, in addition, you had to keep your engine throttled up, and you had gears to shift—six speeds forward and two in reverse. There was still another control to raise and lower the huge blade in front of the machine.

Bill got the lumbering dozer under way. The Chief told him to swing in a big circle on the broad level acres of the Proving Grounds. Off they went, roaring and rumbling along like a fire-breathing monster as the steel cleats on the caterpillar treads clattered along and bit into the ground. Half-way around, the instructor told Bill to stop the "cat" and drop the twelve-foot blade. This was done by moving a lever just back of the driver's seat. It looked something like the tiller on an outboard motorboat. Bill lowered the blade and started the "cat" once more.

Before he had gone many feet, there was a young mountain of soil piled up against the pusher blade. But the "cat" plowed calmly along, pushing the tremendous load ahead of it. Now they were coming to a slight rise. Bill gave the Diesel more throttle and put his hand on the clutch lever.

"Hey!" the Chief bellowed, over the roar of the engine. "Don't slip your clutch! Shift to a lower gear to handle that load!"

Bill shifted, and the "cat" walked right up the incline with its load. He made a half-turn, raised the blade, reversed the bulldozer, and left his load of dirt. Then, turning in a tight circle on his starboard track, Bill pointed the "cat" back to the starting line again.

"O.K.—cut your engine," the instructor said.

Bill did, and they both jumped down off the dozer.

That was great stuff. Bill was as excited as he had been the first time he had ever driven a car—only *more* so. What a piece of equipment that Caterpillar was.

He joined the other boys while the Chief took a second Seabee over the Proving Grounds course.

"What's it like, Scott?" one of his shipmates asked.

"It's 4.0, and then some." Bill exclaimed, beaming. "But it jars the insides out of a guy. You really need a kidney belt or something when you jockey one of those contraptions over the ground."

As their Proving Grounds work went on, day after day, the boys found that, although taming the "cat" was hard work, it was more fun than anything they had ever done before. The Chief had been right. Their *wasn't* anything that dozer wouldn't do.

One whole day Bill worked shoving dirt up into a big mound that must have covered half an acre. Other Seabees before him had started the mound, and it was his job to make it higher. The top was already about fifty feet high; but the "cat" would walk up it without even hesitating, pushing a load of dirt that would have filled a big dump truck.

Later on, when the mound was finished, the boys leveled it off again with their bulldozers.

They they practiced cutting grades to dimension—slopes and "blue tops"—building fills . . . cutting roads around hills . . . and leveling runways. They even found that, by changing the angle of the blade, they could cut ditches with the "cat."

At first, when pushing dirt over a fill with his dozer, Bill was a little jittery each time he approached the edge. It always seemed as though the whole machine would go toppling down the steep bank. This was because he had to drive the "cat" far enough so that the blade and part of the tracks actually hung in the air, *beyond* the edge. Then the load of dirt would go sliding down the forty-five-degree slope.

The instructor noticed that Bill always inched his bulldozer to the edge of the bank very cautiously, and he guessed what the trouble was.

"You've got to give 'er the gun and plow right ahead, Scott!" he shouted one morning.

Bill cut his engine so he could hear what the Chief had to say.

"Even if you *did* go over the bank, you'd have nothing to worry about. That dozer can walk up or down a straight wall. Well," he admitted, *"almost."*

He climbed up into the driver's seat beside Bill and deliberately nosed the bulldozer down the steep bank which Bill had been timid about approaching too closely. The huge lumbering machine clambered down the steep angle under perfect control, then turned and walked right up again.

"See? There's nothing to it." the Chief said as he handed the controls over to the Seabee again.

He told Bill about an experience he had had when he was operating a "cat" up in the Aleutians.

"I was cutting a roadway halfway up one of those frozen mountains when I ran into a sheet of ice, near the edge," he explained.

He told Bill that the bulldozer slewed around, slid off the edge, and rumbled down the 180-foot cliff—which was very steep.

"Luckily we didn't turn any somersaults," the Chief said. "I got a broken leg out of it, when I was thrown against the controls. But since then I've had plenty of confidence in what one of these babies will do."

After that Bill tried running the Caterpillar down the slope and back, once, just to overcome his dear of *accidentally* overshooting his mark sometime. It worked wonderfully well, just as it had when the Chief did it.

Gradually he gained complete confidence in himself and his machine, until operating it became almost as automatic as driving a car. And as he put the bulldozer through its paces, day after day, he began to understand why the Seabees had been able to perform such miracles, in so little time, when they landed on some "Island X."

NEATEST TRICK OF THE WEEK

"What's the matter, Nick?" asked Bill, "Why so glum?"

Nick Torrio was sitting on the edge of his bunk, reading his technical-training schedule.

"Well, why wouldn't I be?" he grumbled. "Look—they've put me down for the Seabee Pontoon School. I wanted to be a construction man. *Anybody* can lay boards across a string of big rubber doughnuts and make a pontoon bridge."

Bill slapped his shipmate on the back.

"What a surprise you've got coming." he said. "Come on—I'm due over at the Proving Grounds. Let's get under way."

Bill was right. The first day in the Pontoon School classroom Nick Torrio learned two things he hadn't known before.

First, Seabee pontoons weren't "rubber doughnuts" at all. They were big sheet-steel boxes, seven feet long, five feet wide, and five feet deep. Each pontoon weighed more than a ton.

Second, he learned that being a pontoon man was one of the most important assignments a Seabee could be given—and the men who were chosen for this job were considered the best construction experts in the Battalion.

In a way the pontoon classroom reminded Nick of a kindergarten for grownups, for here were a lot of full-grown Navy men playing with what looked like children's building blocks. The "blocks" were really small models of the pontoons. The instructor explained their construction and showed them the "jewelry"—the steel plates and angle irons with which they could be bolted together. Each pontoon had a built-in steel ring, in each corner, so it could be hoisted by a crane.

"This famous pontoon was invented by Captain John Laycock, of the Navy," the chief petty officer told the boys. "And believe you me, it's the neatest trick of the week. This steel box has a thousand and one uses, and we're constantly finding still more ways to use it. Somebody nicknamed this pontoon the 'Jeep of the Seabees,' and the name fits it like a glove."

Demonstrating with small models, he showed the boys how pontoons could be strung together to make a bridge from ship to shore, so that trucks and equipment could be put ashore. They could be used to build a bridge across a river, too—or as a wharf, so that a ship could come up alongside and unload. Or, if a diving or salvage job had to be done, you could build a big pontoon float, run a crane onto it, and tow it wherever you wanted to go.

But the most amazing use of the pontoon was this: you bolt a number of pontoons together in a certain way, attach a huge outboard motor, and—*presto magic!*—you have a barge that can go anywhere you want it to, under its own power.

"Well," Nick thought to himself, "this pontoon job is great stuff, after all."

He was no longer disappointed over his assignment. In fact, the more he learned about pontoons the more excited he became—especially when the instructor told the class the story of how Seabee pontoons saved the day when the Allies landed in Sicily.

"Ordinarily an LST rams its bow right up onto the beach. Then it lowers its ramp, and the tanks and machines rumble ashore," the Chief explained. "But the Navy ran into a tough problem off Sicily. The beach sloped so gradually into the Mediterranean that the LST's would run aground when they were several hundred feet from the shore—and the water would still be six feet deep at the end of their ramps."

All the boys knew that you couldn't expect to drive tanks and Jeeps and bulldozers ashore through water six feet deep.

"But the Seabees had the answer, as usual," the instructor continued. "In almost less time than it takes to tell it, they laid down strings of pontoons from the LST's toward the beach. The LST's threw down their ramps, and the tanks and trucks and bulldozers rumbled out onto the pontoons. By the time they got to the end of the pontoon causeway, they were within 'wading distance' of the shore. The enemy didn't know we had any tricks like this up our sleeves, and so they weren't expecting us to land there. We took them by surprise—and you know what happened then."

As Nick learned about the many amazing things that could be done with Seabee pontoons, he made up his mind that he would try to think up still other ways in which they could be used.

After many days spent in the classroom, working with models and watching movies of pontoon operations, the boys finally went down to the harbor to practice working with real pontoons. They learned how to assemble pontoons in various combinations for different purposes. As the crane would swing around and lower each pontoon into place, the Seabees would slip thc bolts in, tighten them, and get set for the next one.

One morning they built a pontoon barge. The crane lowered the pontoons into position, one after another, until the boys had bolted them all together in a combination which was three pontoons wide and seven pontoons long.

The crane's last load was the giant outboard motor, known as a propulsion unit. This, too, was very much like one of the regular pontoons. But it was bigger, and was equipped with a Chrysler engine, a propeller, a steering wheel, and a rudder.

As soon as the propulsion unit had been bolted securely to the other pontoons, the crane lifted the whole affair into the air and set it down on the water. The Seabees scrambled aboard and started the engine, and off they went across the harbor on their fifty-ton Navy barge.

Barges such as this could be used for many different jobs. In the South Pacific, the instructor told them, Seabee barges were often used to bring drums of aviation gasoline and other supplies ashore from ships which remained anchored out beyond the coral reefs. And then, if you needed the pontoons for some other job, you simply unbolted them and strung them together again in some new combination.

The pontoons could be bolted together in such a way as to make a seaplane ramp, if need be. They could also be made into dry docks for repairing ships. This is how the Seabees built a pontoon dry dock . . . .

First they constructed a big pontoon barge, by bolting forty-eight of the compartments together. Then they bolted additional pontoons along both sides of the barge, in an upright position. Opening the valves in the barge pontoons, they let water in—just enough to sink the barge below the surface. When all is ready, a PT boat came up into the harbor and slowly nosed its way onto the barge, between the upright pontoons. The Seabees then forced the water out of the barge with compressed air. Gradually it rose again until it was floating on the water. And there sat the eighty-foot PT boat, in its pontoon cradle—high and dry, ready for repairs.

As their training continued, down on the water front, the Seabees became expert at rigging up pontoon causeways for unloading heavy rolling equipment from landing ships. One day they beat all records by stringing their pontoons from an LST to the shore in five minutes flat. The LST had scarcely grounded when trucks and bulldozers were chugging off the ship toward the shore, several hundred feet away.

"Good work, men." was the Chief's comment. "In a landing problem like that, every second counts—especially when Zeros or Messer-schmitts are strafing the landing party."

By this time Nick Torrio was certain that he had one of the best jobs in the whole battlion. Not for any amount of money would he have traded his chance to be a "bronco-buster" riding a string of Seabee pontoons ashore on "Island X."

KEEP THE HOOK MOVING

"Sometimes this Station reminds me of a dozen three-ring circuses all lumped together." Bill Scott remarked one evening after mess.

"It sure does," Tex Rogers agreed. "We're all so busy, learning the special jobs we've been assigned to, we never get a chance to see what's cooking with the rest of the gang in the Battalion."

But there were two things they knew for certain. There was plenty "cooking" all over the Station—and it wouldn't be long, now, before the 125th Battlion completed its training and shoved off for "Island X."

Some of the boys were spending almost all their time, these days, in the new Stevedore School. They were being trained to do just one important job—that of loading and unloading cargo ships.

"When we first got into the war," the instructor explained, "one of the worst problems was that of unloading cargo ships filled with supplies for our fighting men. Vessels waiting to be unloaded used to clutter up the harbors at some of our overseas bases, and what helpless targets they were for enemy bombers. More than one cargo ship, filled with equipment which American workers had spent months to build, was sent to the bottom simply because the enemy bombed it before it could be unloaded."

The chief petty officer also told the boys that very often freighters and cargo ships had to depend upon native laborers to do the work. Sometimes they even had to ask soldiers and Air Force ground crews to pitch in and help.

"But now, thanks to the Seabee stevedores, that dangerous bottleneck has been broken," the Chief said.

There was a cutaway model of a Liberty ship in the classroom, almost fourteen feet long. By studying it the boys learned where the different compartments were located. Using small models of trucks, tanks, and crates, they practiced the proper methods of stowing cargo in the ship's holds.

Down near the dock area, at the edge of the harbor, a full-size model of a Liberty ship had been built. On this land-locked ship there were the same holds, compartments, and loading gear that the Seabees would have to work with when they reached the battle zones and began unloading real cargo ships.

Day and night, in all kinds of weather, the stevedores practiced loading and unloading cargoes. They knew that supplies must be kept moving regardless of sleet, hail, rain, or broiling sun.

Over and over again they hoisted crates and boxes and heavy machinery up over the sides of the ship. And when they got the ship fully loaded, they unloaded her again.

Up went the cargo nets, loaded with bags, Jeeps, ammunition boxes, trucks, and food. The watchful eye of a chief petty officer checked each operation. The dockside rang with the clatter of block and tackle, the grind of winches, and the bustle of tractors hauling up a never-ending stream of supplies.

The efficiency of any stevedoring job depends upon the speed with which the cargo hook hoists the big nets aboard. So the Seabee stevedores had a battle cry of their own. Hour after hour you could hear it being shouted up and down the dock: "Keep the Hook moving! Keep the Hook moving!"

The Seabees became so skilled as stevedores that they were soon able to trim hours and even days off the time it ordinarily used to take to unload a cargo ship.

Other groups of Seabees were taught how to pipe gasoline and Diesel oil from tankers, anchored offshore, to storage tanks on land. It was quite a trick to rig a long fuel pipe from the shore out to a ship, but it was a method they knew they might have to use over and over again on "Island X."

The eight-inch pipe line for gasoline came in standard sections, thirty feet long. Using flexible couplings, the boys would fasten the sections of pipe together, seal the end with a cap, and shove it out into the water in the direction of the ship as fast as they could build it. The pipe line would float, because it was full of air. When it had been built long enough so that it reached the tanker, the men on the ship would haul it aboard, couple it up to their gasoline tanks, and start pumping the aviation fuel ashore. Sometimes so many sections had to be joined together that the pipe line would be six hundred or seven hundred feet long.

Some of the biggest Quonset huts on the Station were set aside as machine shops, engine schools, and forge shops. Here the boys not only learned how to repair all the different kinds of machinery used by the Seabees but often had to make new parts that were needed.

Among other things they had to tear down a Jeep, to the last nut and bearing, and then assemble it again. In order to give them practice in repair work, all the trucks, station wagons, and staff cars on the Station were brought to this school when they needed fixing up.

"Sometimes, when your Battalion is overseas, certain spare parts you need will be thousands of miles away," the instructor reminded the Seabees. "What do you do in that case—just sit down and wait? You do not. You get busy and make the part you need."

So at the school, even though the needed spare part might be sitting on its shelf in the stock room, the boys sometimes *made* the new crankshaft, gear, or bearing they needed. They became so good at this sort of thing that they could have built a brand-new Jeep, starting from scratch, just by making a duplicate copy of each part contained in the car, from the engine itself to the rear axle.

Some of the Seabees became communications experts, working with both radio and telephone systems. They learned to set up and operate telephone switchboards and field units, and to string their telephone wires wherever they had to go—often under fire.

"When we kicked the Japanese out of the Aleutians," the instructor said, "our communications outfit set up a telephone system that would have cost a real sum of money back in the States. And they did it under the most difficult conditions—in spite of fierce Arctic williwaws, blinding snowstorms, and temperatures that sent the thermometer scuttling to fifty below zero."

In the South Pacific, he added, telephone linemen found that hundred-foot palm trees made the finest telephone poles in the world.

There was a special Fire Fighting School at Camp Endicott, too. In their classroom, and out in the field, the Seabee fire fighters learned how to operate and maintain the various types of equipment their Battalion would take with it overseas.

One of the stunts they had to do would have gladdened the heart of any fire-chaser. First they dug a pit about fifteen feet in diameter. Next they poured several drums of fuel oil into the pit and set it afire. Smoke and flame billowed a hundred feet in the air. And then the Seabees swung into action, trotting out a Chrysler high-pressure fire-fighting unit mounted on wheels. They pointed the long spray nozzle at the flaming oil. Dense chemical fog shot out of the nozzle under pressure, completely smothering the roaring fire in fifteen seconds.

Many of the Seabees were being trained as carpenters. By studying maps of all the overseas bases that had been built thus far, they learned how to lay out a typical Seabee base—where to build the fuel and water tanks . . . the best location for the Quonset huts, repair shops, and other the other buildings that had to be thrown up when a Battalion went ashore.

The most surprising thing, at least to many of them, was the hundred-and-one uses to which a carpenter's common steel square could be put. It was not the simple tool most of the boys had always thought it to be. For example, with a carpenter's square you could determine the pitch of a roof, find the center of a circle or the unknown side of a triangle. You could even lay out that most most difficult of patterns—a star.

There was a special school for Seabee riggers, too. Here the boys were taught how to tie the ten basic knots, how to splice ropes and steel cables, and how to set up a derrick for hoisting multi-ton loads. They also became skilled in rigging a "Chicago boom"—the jungle equivalent of a crane—using just a block-and-tackle and a tree.

Everywhere you went, all over the Station, you could see groups of Seabees learning the jobs which would make them valuable members of their Battalion.

Along the dock you would see them learning to drive piles, using a crane and drop hammer.

And, crisscrossing the harbor, there would be boat loads of the boys practicing the seamanship theories and rules they had learned in the classroom—visual signaling . . . rules of the road . . . how to approach and secure to a mooring buoy . . . boat handling and maneuvering . . . proper use of the compass . . . how to take bearings . . . and all the hundreds of other things a good seaman has to know.

If you wondered back from the harbor, you would run across groups of Seabees learning to become map makers. Some of them would be using standard military mapping kits, with all the most modern instruments. Others would be working out a more difficult assignment.

The job of this second group was to go out and make a map of a certain area, using only a ruler, a piece of string, and some small pieces of scrap lumber. They knew that the instructor already had an accurate detailed map of this same area, and that he would check their work against this map when they got through.

Their first task was to build a crude tripod for surveying. Then they made their measurements, elevations, and scales, and drew up their diagrams. With all this information worked out, they then went back to the classroom and made a careful map of the area. In almost every case the maps made with these crude implements were accurate—even as accurate as those made with the standard military mapping kits.

Over in the rock quarries that were a part of Camp Endicott there was a training program for the explosive experts. In this area the Seabees learned how to use a rock drill for boring holes into which they could place their charges of dynamite. They studied all the different types of explosives they would have to use, and how to detonate them. Some of them were set off electrically, by pushing down a plunger. Others had time fuses.

Working alongside the blasters were Seabees operating pneumatic jackhammers, breaking huge slabs of rock into smaller chunks that could be handled in the big rock-crushing machines.

There were also classes in mosquito control . . . sheet-metal work . . . purifying water for drinking purposes . . . refrigeration . . . erection of wood and steel storage tanks . . . air compressors . . . electric motors . . . generators . . . concrete construction . . . dam building . . . drilling wells . . . photography . . . pumps of all kinds . . . soil testing . . . lubrication . . . plumbing . . . laying metal mats for airplane runways . . . paint sprayers . . . riveting . . . power shovels . . . road scrappers . . . and hundreds of other projects.

Gradually the Battalion was being welded into a magnificent group of workers and fighters—"Jacks-of-all-trades," and every Jack an expert at his trade.

GOOD-BY TO CAMP ENDICOTT

The Seabees' training was over. Bill Scott, Tex Rogers, and Slim Brown were in their barracks, passing the latest scuttlebutt back and forth, when the door suddenly burst open. In dashed Nick Torrio, waving a sheet of paper in his hand.

"Hey, look! I d-did it!" he stammered excitedly.

He held the paper up for them to read.

"Sure, we *all* did it," Bill said.

And he, too, flourished a diploma like Nick's.

"Yeah, I know," Nick persisted. "But look at the mark I got."

Bill read the diploma.

Certificate of Completion," it went. "*Torrio, Nicholas, has this day completed the advanced course in the N.C.T.C Pontoon School with a mark of 4.0.*"

It was signed by the chief instructor.

"Nice going, Torrio, Nicholas." Bill congratulated him.

"Tex got a 4.0, too. Slim and I settled for a 3.8."

"Bill and I both have good alibis, though," Slim said, winking at Tex. "Bill would have copped a 4.0 if he hadn't driven his "cat" over the Chief's foot one day."

"Sure I would have." Bill grinned. "And you queered your chances for a 4.0 the day you made that wisecrack and the instructor in the Camouflage class heard you. Remember? You said that the way he was making you daub your uniform with paint made it look like an explosion in a soup factory."

"Some alibis." exclaimed Nick. "No brains, I'd say."

Tex glanced at his watch.

"Say, we've got to quit shooting the breeze and get under way. The big show goes on at 1100."

They changed into their dress uniforms and hurried out to the parade grounds, where the Battalion was already forming. The "big show" was the presentation of the colors and final inspection.

The ceremony was brief but impressive. As the Battalion stood rigidly at attention, the Commanding Officer of the Station stepped forward and handed the flag, or ensign, to the Commander of the Battalion. On the flag was the legend "125th Battalion." After the inspection the Captain and his aides watched the Battalion pass in review, and the ceremony was over.

At 1700 that afternoon the Station was secured. The Seabees streamed out of the main gate and scrambled aboard the busses for Providence—and home.

After their ten-day embarkation leaves were over—ten days spent visiting friends and family—the Seabees returned to Camp Endicott. Never before had ten days passed so quickly. But, back again with their Battalion, the Seabees were soon too busy to have even a pang of homesickness.

No longer were they billeted in Barracks 54. The entire 125th Battalion was now quartered in the "ABD" area, adjoining the main Station. "ABD" stood for "Advanced Base Depot." This was the jumping-off place, from where they would go to board their ships for "Island X." And they knew they might have to embark on short notice, almost any day now.

"Golly, I can hardly wait." Slim said one morning. "I've seen pictures of those tropical isles. Blue water to swim in . . . beautiful native girls with flowers in their hair to wait on you . . . what a life."

But he was only kidding. He, and all the other Seabees, knew that "Island X" would not be anything like that. It most probably would be a war-torn strip of atoll, a mere pin point in the Pacific, with its palm trees ripped bare by shell fire, its beaches and landing areas pocked by huge bomb craters—and perhaps some very hostile individuals waiting there to welcome them.

In the meantime the Seabees had too much to do to think very much about where they were going, even if they had known. The Battalion was being outfitted now for its over-seas base, wherever it might be. The gigantic task of assembling gear and supplies was under way.Cargo ships and transports were already tied up alongside the docks at the ABD area, which was piled high with crates and boxes and all sorts of equipment—arms and ammunition; clothing, food, medical supplies; office supplies and records; artillery and mortars, shells; and all the tools necessary to build a naval base, from screw drivers to bulldozers, from nails to power shovels.

Day and night the Seabee stevedores labored to get all these supplies into the ships' holds. Finally the colossal task was finished. The ships were loaded, and the Seabees received their orders to embark.

"Well, come on, guys," Bill said. "This is it."

They packed their sea bags, slung them over their shoulders, and swung up the gangplank to board their ship.

The lines were cast off and a few minutes later the 125th Battalion of Seabees was under way, to a mysterious "destination unknown"—unknown except to the captain and commanding officers aboard each ship.

As his ship slid out through the harbor, Bill Scott sauntered back and joined several of his shipmates who were standing at the rail to catch a last glimpse of the Station where they had spent so many weeks of rigorous training. They leaned on the rail, watching while Quonset huts and sprawling acres of Camp Endicott slowly faded away in the distance.

Now they were seasoned, hardened members of a Seabee Battalion—outward bound. They know that they would have to face many hardships before they ever again saw the shores of America. And they well knew that some of them might never return. But, whatever was in store for them, they were ready.

WELCOME TO "ISLAND X"

The time: *three weeks later.*

In the early-morning haze the Seabees caught their first glimpse of "Island X," dead ahead.

"There's your peaceful tropical isle, Slim." remarked Tex Rogers.

Several of the boys were standing by the rail as the convoy moved in. Now the ships, still about a mile offshore, were dropping anchor.

Slim shaded his eyes and scanned the shore line. All the boys were dressed in full battle gear—helmets, carbines slung over their shoulders, and gas masks ready for instant use.

"Funny thing," Bill remarked. "I can't see that beautiful native girl trotting down the beach to meet us with a basket of tropical fruit on her head. Didn't anybody tell her we were coming?"

Several of the Seabees smiled grimly. Beautiful tropical isle, indeed. Even at this distance they could see what was in store for them. The word had already passed around that the Marines had landed, just two days before. And it had been pretty rugged going for them.

Mile-high pillars of black smoke still billowed skyward from blasted Japanese installations in the jungle back of the beach-head. The beach itself was dotted with wrecked artillery pieces, Jeeps, and half-track tank busters. Several half-sunken barges gave mute evidence of the fierce reception the Marines had met with. The palm trees for hundreds of yards back were stripped and torn to shreds, as though a hurricane of explosives had ripped through them. Even now the sound of heavy artillery and the *rat-a-tat-tat* of machine-gun fire back in the hills drifted out over the water.

Suddenly the cargo ship's "squawk box"—its loud speaker amplifier—made a rasping announcement.

"Now hear this! . . . Now hear this!" it blasted out.

The Seabees stiffened and paid close attention.

"The Marines have driven the Japanese four miles back into the jungle and are mopping up, against fierce resistance. Unloading operations will start immediately. The 125th Battalion will land and proceed according to plan."

The "squawk box" repeated the message.

Now a Marine signalman, standing on the beach, was wig-wagging a message to the cargo ships and transports. Even as a Seabee wigwagged an answer, the huge cranes and winches aboard the ships started to unload their cargoes. The first thing to go over the side was a completely assembled Seabee pontoon barge, with its outboard propulsion unit attached. Nick Torrio and the rest of his company had put it together just the day before. Now another one was being hoisted up into the air, and was lowered slowly into the water alongside the ship.

The Seabee stevedores sprang into action. Out of the ships' holds they raised bulldozers, boxes of food and ammunition, power shovels, mobile four-inch guns, shells, and all the other construction equipment and armament they could crowd onto the decks of the fifty-ton pontoon barges. Bill Scott saw his big yellow eighteen-ton "cat" swinging up out of the hold, then out over the side of the vessel and down onto the pontoon.

At a given signal he and about twenty other Seabees crawled over the rail and scrambled down the cargo net onto the deck of their barge. Then—suddenly—it happened.

"General quarters! Man battle stations! Japanese fighters and dive bombers coming in—position two o'clock!"

The "squawk box" frantically repeated the alarm.

Seabee gunners pounded across the decks to man their anti-aircraft guns. Bill and his companions on the barges crouched low behind the bulldozers and other heavy equipment for protection.

Then they saw the Nip planes coming in—tiny specks directly in the sun, so that the glare would make it difficult for the Navy gunmen to sight them.

The next minute it seemed as if the end of the world had come. Planes screamed down from the sky on all sides, straffing and bombing the ships. One Japanese aviator swooped down mast high in a power dive, raking the ship with his guns. He passed so close that Bill could see the grinning pilot as he zoomed upward again for altitude.

There were only nine planes, but those nine seemed like a thousand. They were everywhere at once. Bill looked up over the edge of his bulldozer just in time to see a lone Japanese dive bomber streaking down toward his ship. Down . . . down he came, with the speed of a bullet. When he was only about a hundred yards from his target, a Navy gunner got him squarely on the nose. The propeller burst into a thousand fragments; smoke and flame shot from the engine . . . and the shrieking dive bomber plunged into the sea, not fifty yards from Bill's barge. The plane disappeared beneath the surface, leaving a pall of smoke. The next instant there was a rumbling explosion under the water, and the sea was churned up for one hundred and fifty yards around the spot where the plane had vanished.

"There go his bombs." Nick said quietly. "His eggs exploded."

Soon the surface of the water was dotted with dead fish, killed by the terrific subsurface explosion. A wing tip and section of the rudder bobbed to the surface. But there was no sight of the Japanese pilot.

"Some wandering shark will have a feast when his carcass comes loose from the wreckage," Bill commented.

"There goes another one!" Nick shouted above the deafening noise of the gunfire and the screaming Zeros.

Sure enough, another Japanese plane was plunging into the sea—trailing a red and black streamer of flames and smoke. It hit . . . and exploded. A huge geyser of water shot into the air.

"Scratch two Zeros!" Bill yelled excitedly.

Off to the starboard, still a third Japanese plane was in serious trouble. The gunners had scored a direct hit on his engine. With a thin streamer of white smoke trailing behind him, he turned and high-tailed for home.

"Now they're all going." Nick said.

He was right. Out of bombs and ammunition, the six remaining Japanese planes had turned and were streaking back toward their home base.

"That last one will never make it, I bet." Bill said.

He glanced at his watch. The entire attack lasted only five minutes. It seemed like hours. In the heat and excitement of the aerial battle, there had been so much going on that Bill had forgotten to be scared. But now that it was over he realized that he was thoroughly frightened. His hand trembled as he pushed back his steel helmet and mopped his brow.

"Whew." he exclaimed. "What a warm welcome to 'Island X'."

"You know something, Bill?" Nick grinned, now that the color was starting to come back into his face. "I don't think those Japanese pilots *like* us."

Bill looked up the side of their big cargo ship. He saw a corpsman helping a Seabee gunner out of his gun turret. Blood was streaming down the gunner's arm.

Bill turned to Nick and said through gritted teeth, "That goes double, guy. They'll get a dose of their own medicine, and I don't mean maybe."

By some miracle very little damage had been done by the attacking Japanese, and casualties were light. A few wounded Seabees were being patched up, but none had been killed.

The Japanese planes were now two fly-speck V formations over the distant horizon.

"Resume landing operations." the "squawk box" blared out. "Resume landing operations."

The propulsion units on the two pontoon barges roared to life, and the heavily loaded floats started their one-mile voyage to the shell-torn shore of "Island X."

THE SEABEES TAKE OVER

Hour after hour the fleet of Seabee pontoon barges kept ferrying back and forth between the ships and the beach-head, bringing in supplies.

Temporary "headquarters," consisting of half a dozen empty gasoline drums, had already been set up on the beach. Squatting on the drums, several Marine and Naval officers were studying a map of the island and laying out a plan of strategy.

"Things are in pretty bad shape here," said a young Captain of Marines. He pointed to the wharves and buildings of the former Japanese base, now nothing but smoldering heaps of rubble. :Most of the drinking water on the island is contaminated—it's not fit for use. My men have been living in slit trenches and fox holes ever since they landed, and we're getting a little low on supplies and ammunition. In other words—when I say we're glad to see the Seabees moving in, that's putting it mildly."

"We'll soon have the base shipshape, sir," a Seabee lieutenant said. "Aside from the water supply, what's the most urgent problem?"

"There are *two* urgent problems, and they both come first." the Marine officer replied, with a grin. He pointed to a spot on the map. "My men are back here, four to nine miles from this beach. We need some sort of road slashed out of the jungle so we can haul supplies in. The other problem is—well, you can see it for yourself." He swept his arm over the area which used to be an airplane landing strip. "We need air protection, but we can't get our planes in until we have a place for them to set down."

The group of officers walked over to inspect the cleared area which was formerly the Japanese runway. Huge bomb craters, made by our carrier-based planes, had chewed the strip to bits. Only the fact that there were no trees gave any hint that it used to be runway. Half a dozen Japanese planes were scattered over the area, completely wrecked—planes which had been strafed on the ground before they had a chance to take off.

"It wasn't a very good strip to begin with," one of the Marine officers said. "Like most Japanese strips, it was too short—only thirty-seven hundred feet long. And it was built so poorly that we wouldn't ask a Marine pilot to land a wheelbarrow on it, except in an emergency. What we need now is a hard-packed runway eight thousand feet long, and three times as wide as this one."

He pointed to a grove of shattered palm trees and a hill, just beyond the bomb-pocked strip.

"Those trees will have to come out and the hill will have to be leveled off. Can you boys do it for us? We need air support, and we need it badly. The Japanese keep flying over several times a day."

"Sir, your air strip is practically built right now." the Seabee Officer in Charge exclaimed.

He jogged down the beach to the spot where half a dozen bulldozers were fueling up.

"All right, boys." he said. "You've got a job cut out for you."

On the map he showed them just where the Marines had dug in, back in the jungle. Tracing a rough route on the map, he told two of the bulldozer operators to slash a roadway from the beach through the dense undergrowth to the Marines' position.

He turned to the other bulldozer operators.

"You—Scott, Jacowski, Butler—knock out an air strip for the Marines on the double. That palm grove comes down. And that hill has to be chopped off, too."

He indicated the size which the new air strip should be when completed.

The boys started their giant bulldozers and went clattering and snorting up the beach to build the jungle road and the airfield.

In the meantime there was plenty of work for every Seabee specialist to do. Some were hastily throwing up a temporary tent city, with a slit trench beside each tent in case of an air attack.

Seabee electricians were setting up large generators to provide light and power. Seabee carpenters were knocking together machine repair shops, and building wooden forms for concrete gun emplacements.

The concrete mixers were already chugging away, churning the mixture which would be poured into the frames.

"Nail 'em up, boys." the operators shouted to the carpenters. "Here comes the first batch."

Other Seabee detachments were setting up huge stills. They promised that within a few hours their stills would start delivering hundreds of gallons of fresh water an hour, distilled from salt water pumped up from the sea.

Kitchen and cooking equipment was coming off the pontoon barges now, for it would soon be time for the first "chow line."

Well drillers were busy, driving down into the earth for a permanent water supply for the new Naval base. Tank builders were erecting huge steel storage tanks for gasoline, water and Diesel oil. Munition dumps, back some distance from the beachhead, were being built.

"Gangway!" yelled the tractor drivers as they started to drag up boxes of ammunition and shells from the barges on the wooden skids.

As fast as each installation was completed, the camouflage experts got busy and hid it from view.

Over on the air strip the palm trees came crashing down like pins in a bowling alley before the blade of Bill's giant yellow "cat." The other bulldozers were filling in the bomb craters and leveling off the hill at the end of the runway the Japanese had laid down.

Above the roar of the machinery the noise of gunfire, back in the jungle, could be plainly heard every now and the.

Suddenly an air-raid warning rang out. The boys jumped down off their machines and "hit the deck." Bill tumbled into a near-by crater for cover. The Seabees who were busy on their installations scurried into the nearest slit trench.

Then the Japanese bombers came over, flying high. They dropped their loads, then turned and made another run over the base to observe the results. The ground shook as if from an earthquake, at the conclusion of the bomb hits. Most of the bombs fell into open areas, doing little damage but showering dirt and debris high into the air. Five minutes later the "All Clear" sounded, and the work of constructing the new base continued as though nothing had happened.

Later that day a Seabee electrician by the name of Collier performed an act of heroism which won him a commendation from the Commanding Officer. It happened this way.

A detachment of Seabees was emplacing a 40-millimeter gun in a hillside, just beyond the camp, when a Japanese sniper began to annoy them. The Japanese soldier had sneaked through the outposts and was concealed in a cave beyond a swift stream, about three hundred yards away. Collier was stringing a communication line into the emplacement when the first bullet whined by, just over his head.

"Go ahead with the job, gang," he said, throwing down his pliers and unslinging his carbine. "I'll get that Jap!"

Collier made his way toward the stream, crouching behind rocks and trees as he went. The sniper saw him. Every time the Seabee left the cover of a rock to gain another few yards, the sniper took a pot shot at him. Finally Collier reached a rock right at the edge of the stream. He knew that the instant he stood up to cross the stream the sniper would fire. But he had an answer for that one. He waited patiently for what seemed an hour. Actually it was only a few seconds. Then, as the sniper popped his head up to see what Collier was doing, the Seabee let him have it. The Japanese soldier never had time to squeeze the trigger.

Collier crossed the stream and scrambled on up to the cave. Hastily he searched the victim for the cocked grenade which even a dead Japanese soldier so often conceals in his blouse. He found it and heaved it into the stream before it could explode.

"O.K., fellows." he yelled.

In another five minutes he was back at the gun emplacement, calmly finishing the job of stringing his wire.

When darkness fell, at the end of the first day, the Seabees set up floodlights and kept right on working. That Air strip *had* to be completed. Except for refueling the bulldozers and grabbing a bite to eat at chow time, Bill and his shipmates never quit working—for twenty-four hours at a stretch.

Seabee carpenters were now replacing the tent city with portable Quonset huts for Marine and Navy personnel. A permanent mess hall was almost completed, too.—and a sick bay, fully equipped and staffed by the skilled medical officers who had come out with the 125th Battalion.

At 1200 on the second day, in heat that broiled the men and made metal objects almost too hot to touch, three things happened.

One—the ice machines had frozen several tons of ice, and the Seabees had real ice cream for dessert.

Two—the fighting Marines came back from the other side of the island. Except for a hundred or so Japanese stragglers, whom it might take weeks to mop up, their job was finished.

Three—the air strip was completed, and the first flight of Marine fighter planes came in.

"Island X" was now fast becoming a full-fledged U.S. Naval base—one more stepping stone on the relentless march toward Tokyo.

In another two weeks it bore little resemblance to the crude Japanese base which the Marines had stormed and taken. The airplane runway was now built of hard-packed coral, scooped up from the reefs offshore and crushed firmly down. Even the Flying Fortresses and Liberators could land on the field now. Nine of these mammoth four-engine bombers were already snugged down in camouflaged revetments beside the runway.

Permanent buildings were almost completed. There was even a recreation hall—and beyond it a baseball diamond.

Heavy trucks loaded with supplies—and Jeeps too—rumbled along the fine roads which the Seabees had built on "Island X."

Docks and wharves had been built out into the lagoon, so that cargo ships and transports would come in to unload their supplies and men.

One of the ten-mile highways that skirted the beach of "Island X" had been christened "Marine Drive." The Seabees had put up a sign which read:

TO OUR GOOD FRIENDS AND ABLE PROTECTORS, THE FIGHTING MARINES, WE DEDICATE THIS HIGHWAY

Not to be outdone, the Marines had erected another sign a few hundred yards farther along. This sign read:

WHEN WE REACH THE ISLE OF JAPAN, WITH OUR CAPS AT A JAUNTY TILT, WE WILL ENTER THE CITY OF TOKYO ON THE ROADS THE SEABEES BUILT!

"BOZO" WINS HIS RATING

Life on "Island X" gradually settled down into daily routine activities.

Now that there was a bomber group and a fighter squadron based on the island, Japanese air raids were no longer the threat they had been in the beginning. Enemy planes still came over, to be sure. But they always met with a hot reception.

One day nine Japanese bombers flew over. They came in high, at about thirty thousand feet. From past experience they had gained a healthy respect for the anti-aircraft defenses of "Island X" and the fighter planes that invariably took to the sky to drive them off.

Bill Scott was working on his bulldozer when the air raid alarm sounded. He had spent the past hour cleaning the air filter and making minor adjustments on his Diesel engine. Now, in memory of the home-built tractor he had left behind, he was lettering a name on his "cat." He had just finished painting the "G" on YELLOW BUG when he heard the drone of the enemy bombers. He set down his paint can and scrambled into the slit trench to watch the show.

The P-38's were already warming up on the ramps. Now they roared down the runway and took off. High overhead—mere specks in the cloudless blue sky—came V formations of enemy bombers. When they saw the American pursuit planes taking after them, the formations broke up; and the planes scurried for safety, dropping their eggs harmlessly in the dense jungle.

"He got him!" Bill shouted as he saw a P-38 Lightning riding an enemy bomber's tail and shooting a trail of hot lead into the plane.

The Japanese bomber burst into a puff of black smoke and went into a spin, out of control. It plummeted to the earth, burning like a sulphur match.

Another P-38 scored a direct hit on a bomber which had not had time to release its bombs. The entire plane exploded in mid-air. One minute it was there—and a second later there just wasn't any of it left.

In a few minutes the sky battle was over. The victorious P-38's came back to the landing strip. The score: three Japanese bombers destroyed and two damaged. All our planes returned safely to their base.

In between air raids and construction jobs the Seabees had a little time to amuse themselves—swimming in the blue lagoon, playing baseball with the Marines, or just plain "shooting the breeze." Some of the boys had caught bright-plumaged parrots that lived in the jungle, and were making pets of them. So far none of the parrots was willing to talk, but the Seabees patiently kept on trying to teach them to say words.

The favorite pet on the base was a little monkey whose curiosity had betrayed him. The night he decided to give up jungle life for the civilized way of living on a Seabee base, he nearly caused a riot in the barracks. He kept swinging from bunk to rafter, and from rafter to bunk, always making a perfect three-point landing.

He was the friendliest little beast, and after he had convinced the startled Seabees that he was just a plain monkey, and not a Japanese spy, he got along famously. In fact, he was so friendly that Bill Scott suggested giving him a name and a Navy rating.

"Let's call him Bozo," Bill said, "with a rating of EF 1/c."

"Bozo is a good name for the monk," Tex agreed. "But what does the EF 1/c mean?"

"That stands for 'Everybody's Friend, first class," Bill replied.

The monkey soon got the habit of following the boys wherever they went. Much of the time he spent on the seat of Bill's bulldozer. The roar of the engine, instead of frightening him, seemed to fascinate him. Sometimes he annoyed Bill by grabbing the controls with his little brown fingers, or by trying to shift gears.

One afternoon, after several hours of road grading with the "help" of Bozo, Bill stormed into the barracks with the monkey on his shoulder. Bill was chattering away furiously.

"Here." Bill exclaimed in disgust. "Somebody else can adopt this darn monk. I'm through with him."

Tex held up his arms. Bozo made a flying leap through the air.

"I guess we'd better change your Navy rating, Bozo," Tex said. "From now on you're busted to Bozo, GNNC."

Bill laughed. "What's that mean?"

"General Nuisance, No Class." Tex replied.

In spite of Bozo's troublemaking on the bulldozer, Bill couldn't stay angry at the monkey very long. All the boys got a big kick out of the way he would take a Coca-Cola bottle in his eager little hands and drain it right down to the last drop. But it wasn't quite so funny when he clambered up to the rafters and dropped empty "Coke" bottles on the unsuspecting heads of sleeping Seabees.

That wasn't all he did, either. Bozo took great delight in spilling things on clean bedding and upsetting anything that *could* be upset. Then, before he could be caught and punished for his prank, Bozo would dash outside and scuttle up a palm tree—and sit, chattering defiance at his masters.

Finally the Seabees lost all patience with the monkey. To cure him of his antics they built a "brig," where he was kept confined except for certain periods. It did no good to chase him back into the jungle. He kept coming back, preferring even the "brig" to the loss of his good Seabee friends.

Monkeys and parrots were not the only wild life on "Island X," by any means. One day a certain chief petty officer found that still other creatures claimed the right to live on the new Naval base.

The officer was in charge of a detail of Seabees who were stringing telephone wire from one tree to another. Suddenly he found his men staring uneasily up into a tree making no attempt to take the wire up.

"What's the trouble?" he asked. "Let's get on with the job."

A Seabee pointed up to one of the lower branches of the tree. The Chief looked. There was a three-foot giant lizard, of the most terrifying appearance, lying at full length on the limb.

The Seabees looked at the chief petty officer. The chief petty officer looked at the Seabees.

Then, without saying a word, the Chief fastened his climbing belt to encircle the trunk of the tree, buckled on his climbing hooks, and stated up.

As he slithered up the tree, trying his best to overcome a prickly sensation, the lizard started to climb, too. Every four or five feet the loathsome monster would pause and turn around to stare at the chief petty officer. Then, apparently impressed by the officer's rank, the lizard would start climbing again. Each time he stopped, the Chief would stop, too, and stare back at the lizard.

At a height of twenty-five feet the officer secured his wire and started down. The lizard started to climb down, too—faster and faster.

As the Seabee officer swung to the ground, the men cheered.

The Chief mopped his brow and exclaimed, with a broad grin, "I don't know how fast I was traveling down that tree, but I won."

NEVER A DULL MOMENT

The Seabees were mystified. How did the Japanese stragglers, back in the jungle, manage to keep alive week after week? They had no food, no way of getting more ammunition. But, almost every day, a Seabee or Marine patrol would find several of them skulking through the rank undergrowth. And they were just as full of fight as ever.

Still more astonishing was the way the almost-naked men managed to sneak through the lines to snipe at the Americans.

One morning, just before mess call, a rifleshot crashed into the galley, grazing a Seabee on the arm. The boys made a dash for the slit trenches. And there, plain as day, they could see two Japanese snipers high up in the branches of a palm tree.

They had come in during the night, climbing the smooth eighty-foot trees with the aid of their split-toed shoes. As soon as they saw signs of life about the camp, they got down to business with their rifles. They knew, of course, that they faced certain death. But that didn't seem to worry them, if only they could account for a couple of Americans first.

A Seaman yeoman drew a bead on one of the Japanese snipers. *Ping!* His bullet found its mark. The sniper came hurtling down out of the lofty tree, turning somersaults all the way to the ground.

A Marine got the other one. But the sniper didn't fall out of the tree. His body lay sprawled across one of the branches. So a Seabee lineman had to buckle on his hooks and climb up the palm to dislodge him.

The excitement caused by the snipers had no sooner died down than still other visitors were reported—this time, welcome visitors. The radioman announced that a convoy was standing off "Island X," signaling to come in. Half an hour later the ships appeared—Army cargo ships and transports, escorted by Navy destroyers.

The Marines set up a cheer, for now they knew that the scuttlebutt which had been going the rounds for several days was true. The Army was sending in troops to replace them. This was good news for the war-weary Leathernecks. It meant that they would now receive well-deserved furloughs. For most of them, storming "Island X" had been their third campaign. They looked forward to a rest, and perhaps a chance to go home for a few weeks.

For the Seabees the arrival of the convoy meant hard work, unloading the ships.

As soon as the first freighter tied up alongside the dock, the stevedores pitched in. Up out of the holds came thousands of boxes of ammunition, tons of food, and all the other supplies the Army brought in—Jeeps, tanks, trucks, mobile artillery, caterpillar tractors, and so on.

Luckily no Jap planes came over the island during the unloading operations. It always annoyed the Seabee stevedores to have to interrupt their work because of "Condition Red." They couldn't break records when they had to stop to fight off aerial attacks.

But there was one slight mishap.

A ten-ton "cat" was being hoisted out of the hold. Up . . . up . . . it came. When it had been lowered over the side of the ship and was only two feet above the dock, the hoisting bridle snapped. The tractor hit the edge of the dock and—*kersplash!*—off it slid into the sea. The Seabees looked down into the clear blue water. There sat the tractor, right-side-up on the sand, in about twenty-five feet of water.

An Army colonel peered over the side of the ship.

"A tough break, men." he said. "I guess we'll have to write that tractor off as a loss."

"No, sir!" one of the stevedores shot back. "That tractor cost Uncle Sam five thousand bucks. We'll get it out for you."

Ten minutes later a Seabee pontoon float, fitted out as a diving barge, had nosed its way alongside the dock. Tex Rogers and one of his shipmates, clad only in swimming trunks, were already putting on their diving masks. The barge tied up at the dock, and down went the divers.

Some of the Seabees remained topside. Two of them manned the air pump. Two others acted as tenders, handling the lines and air hoses as Tex and the other diver slipped over the side of the barge.

The men on the dock lowered new hoisting cables from the ship's crane, so that the divers could rig a sling under the tractor.

This was not the first time Tex had walked about on the floor of the harbor off "Island X," but never before had the water seemed so clear. Stripped parrot fish and long-snouted surgeon fish floated fearlessly past his face—as much as to say, "What's cooking, Seabee?"

Then a huge tropical jellyfish, colored all hues of the rainbow, floated aimlessly by as Tex tugged at the steel cable to feed it under the tractor. How white the sand was. And how blue the water. The coral formations reminded him of gorgeous underwater gardens—fantastic flowers in riotous colors, swaying gently with the motion of the current.

Tex always enjoyed working at this depth. He had been down as far as forty feet, with only the diving mask, on other days. Beyond that depth he always wore his helmet, for the pressure became too heavy on his ears. He like shallow diving operations best. Working at only twenty-five feet, there was always the comforting thought that, if anything went wrong, he could quickly slip off his diving mask and rise to the surface.

Finally the new sling was in place. The two divers came up and the barge moved away, so that the tractor could be raised.

The ship's winch started to whir . . . the cables became taut . . . and slowly the "seagoing" tractor rose up out of the water. A few minutes later it was standing high and dry—well, *almost* dry—on the beach. After a thorough going-over by Seabee mechanics it would be as good as new.

Nothing else very exciting happened that day except a Japanese air raid, just at dusk. There were eighteen bombers in the formation, which came in very high. They dropped their eggs without scoring a hit. Our P-38's brought down two of them. The rest got away.

FAREWELL TO "ISLAND X"

Outside Bill Scott's Quonset hut the night was pitch black. The rain was coming down in sheets, pounding on the metal roof in a tropical downpour. The rain had been coming down in torrents like this for three days and nights, without let-up.

Bill sat on the edge of his bunk, writing a letter home.

"Dear Mom—We have been here almost three months now. Things have sort of simmered down lately.

"I wish I could show you a before-and-after picture of 'Island X.' You'd be amazed. Before we came it was a tropical No Man's Land, blasted by shell fire. Today it is one of the finest

Naval bases in the South Pacific—an island empire built by the Seabees.

"There are swell modern hangars, with several squadrons of fighter planes and a heavy-bombardment group made up of Fortresses and Liberators.

"The Marines went away some time ago. There are many thousand U.S. soldiers quartered here now. We call them 'doggies.' Don't ask me why. The Army M.P's make us plenty sore sometimes. They pull us in and report us to the O.D. for speeding in our Jeeps on the highways we built for them. Some nerve.

"Speaking of roads, you should see the job we did here. There's a four-land paved highway we built from one end of the island to the other. It used to be jungle—jungle so thick you could hardly hack a trail through it with a machete. Well, the other day one of our boys decided to clock the traffic on this highway—just for fun. In one hour he counted 873 vehicles whizzing by.

"You can see, Mom, that it's getting quite civilized here. We're hoping to get sent to another spot soon, where there's more excitement. We think it will be -------------- *[censored]*. We heard, just yesterday, that the Marines have landed there—so that means they'll be needing us. If we succeed in taking over *that* Japanese island, it will be one of the biggest victories yet.

"There isn't much more news to write. The other day I was one of a detail sent to an island about twelve miles from here to make a survey for a landing strip. Some of the fellows that went along were the same ones I've been writing to you about in my other letters—Tex Rogers, Slim brown, and Nick Torrio. We had a swell time.

"The natives on these islands don't speak much English, but we manage to get along with them alright. Everything's fine as long as you don't try to kill their pigs. That makes them mad as hornets. They sure hate the Japanese. You'd understand why if you could see some of the things we've seen.

"Well, when we landed on the island, to make a survey, a big guy came down and told us he chief. He had fishbones driven through both ears and there was fresh lime all over his noggin. That's a custom they have down here. The lime is white when they put it on. After a while it turns the hair red. A native with red hair is really *somebody.*

"I told the chief that I was head man of our landing party. We shook hanks very solemnly. Then he wanted to trade or sell us some junk he had—a whale's rib, a native basket, a fish, a club, some shells, a beautifully tanned python skin about ten feet long . . . and, of all things, an umbrella. He traded the python skin to me for ten packages of chewing gum. They love gum. I'll wrap the python skin and send it to you tomorrow. You can hang it in my room—or make yourself some snazzy shoes with it, if you want you.

"The natives were very friendly. They went to work and hacked out a trail for us with their machetes. After we made our survey, we saw another group of natives carrying a giant octopus. He must have weighed all of fifty pounds. He was still alive, too, and making an awful fuss about being carted off.

"The natives would rather eat an octopus than a good roast beef. To catch the critters, they wait until the tide goes out. Then they reach into the little caves under the rocks and feel around. If Mr. Octopus is at home he grabs the native's arm, and then the fun begins.

"The octopus gets awfully mad because he knows it's a sure bet that he'll end up in the stew pot. So he puts up quite an argument with the native, wrapping his long tenacles around his arms, and legs, and body. The native stands there like a dope and waits until the octopus really has a good hold. Then another native rushes up, pokes a spear into the octopus, and unwraps his buddy. The first native is really in a bad spot if his friend doesn't get there in time to stab the octopus."You can see now why we don't cavort and splash around very much in the ocean, Mom.

"Just before we left the island for the return trip, the chief showed us a stained envelope and made signs that meant he wanted us to read the letter that was inside. I opened it up. It was a note written by a U.S. Marines major, saying that Chief Banbawangi was a great warrior.

"Not to be outdone, we wrote a note for the chief, too—one the back of one of our survey forms, with our Battalion insignia on it. Years from now the chief will probably show this note to some missionary or British government official, who will read: 'Chief Banbawangi is strictly a right guy. He extended the hand of welcome to us like a true Rotary Club president. He is of a cheerful disposition—in fact, he's a 4.0 with us Seabees. We will always be glad to put out the Welcome mat and trade a fish with His Honor.'

"Tex and Slim and the rest of us all signed the note with a great flourish, and added our Navy ratings after our names. We have a friend for life."

Bill finished his letter and turned in, just as the bugle was blowing the last notes of "Taps."

For half an hour he lay awake, thinking of home and of his experiences as a Seabee.

Finally he rolled over and whispered, "Tex, are you asleep?"

"Yes, I am." Tex mumbled. "What's the matter?"

"I was just thinking—" Bill whispered back. "It won't be long now before we shove off."

Tex yawned. "That's all right with me," he grunted, still half asleep. "Things are too darn quiet around here."

The boys were right. Two days later a fleet of LST's hove to, off "Island X." The Seabees received orders to load their gear aboard the ships. The Marines had landed on the big island which Bill had mentioned in the letter to his mother—two hundred miles from "Island X." A new base would have to be built.

Once again the stevedores pitched in, loading the tons of supplies and equipment aboard the ships. Finally all was ready and the order was given to embark.

The Commander of the Naval base came out to bid the boys farewell.

"I'm not very good at making speeches," he said as the Seabees stood at attention in their battle gear, "but I'm going to try to tell you what we feel in our hearts about the job you men have done here.

"Through your skill, your devotion to duty and long hours of work, regardless of the danger involved, you have transformed this jungle island into a glowing example of what can be done by the American spirit and the American will-to-win. You have accomplished wonders."

"And you have been good shipmates. You have lived up to the best and finest traditions of the Navy. We hate to see you go. But those of us you are leaving behind have the satisfaction of knowing that you will do a good job on this important assignment you are now going to fulfill. If it is a difficult job, you will do it with the speed which has made you famous. If it is an impossible job, it will take you a little longer.

"There is one flag-signal hoist in the Navy which every fighting man would willingly die in order to deserve. That signal is 'Well done!' To you officers and men of the 125th Seabee Battalion, we now hoist the signal 'Well done!' You deserve it. Good-by—and good luck!"

Half an hour latter the LST's slid out of the blue harbor into a brilliant red and gold sunset. Their foaming wakes spelled farewell to "Island X." ▬

Seabee trainees scaling cargo nets. USN Photo (SEABEE/Lent)

Commando course located at Camp Endicott. USN Photo (Lent)

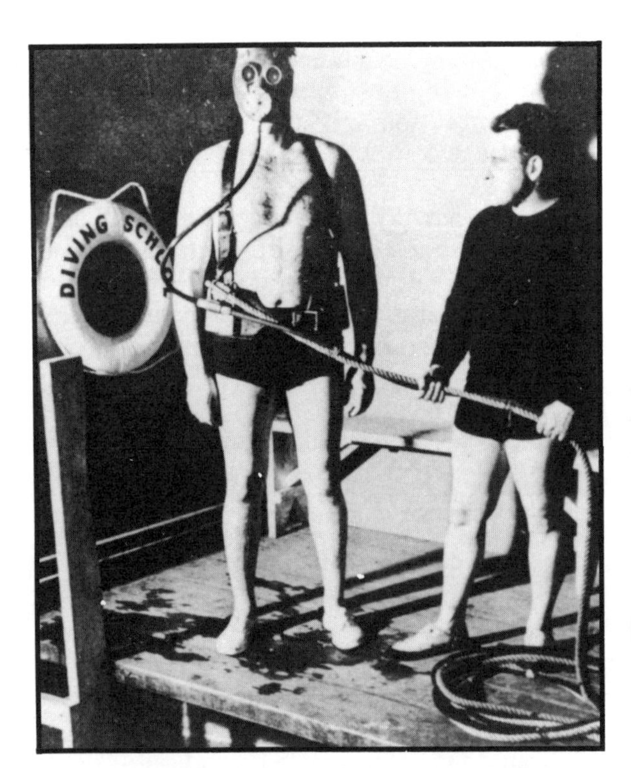

A Seabee trainee, wearing the obsolete-appearing Orco diving mask for shallow water work, as he prepares to work on tasks in the underwater tank. USN Photo (SEABEE/Lent)

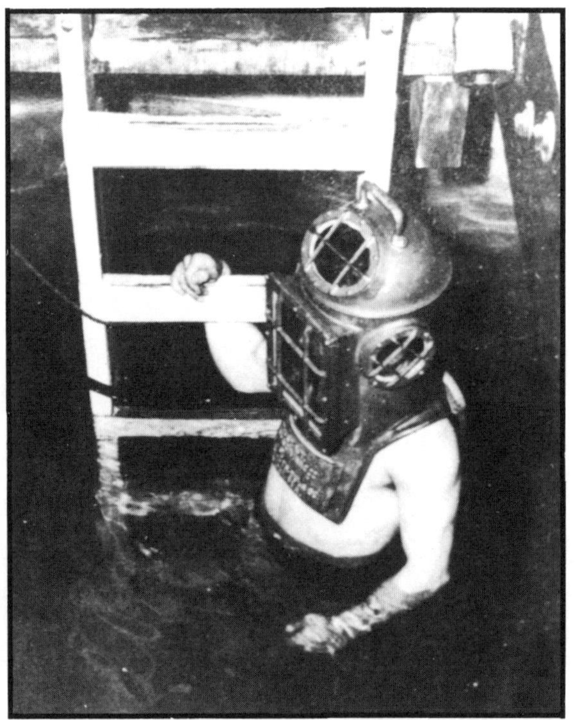

Seabee trainee wearing a deep water diving hood as he descends the ladder into the diving tank. USN Photo (SEABEE/Lent)

Quonset hut at Seabee training base.
USN Photo (SEABEE/Lent)

Mail call at Seabee training base.
USN Photo (SEABEE/Lent)

SEABEE bulldozer operator.
USN Photo (SEABEE/Lent)

Judo training at Camp Endicott
USN Photo (SEABEE/Lent)

Commando course at Camp Endicott. USN Photo (Lent)

Learning to search for concealed mines and booby traps. USN Photo (SEABEE/Lent)

Seabee at the helm of a self-propelled pontoon barge. USN Photo (SEABEE/Lent)

THE BATTALION passes in review after presentation of the colors. USN Photo (Lent)

Judo-boxing at the training base ring. USN Photo (Lent)

Training base library attended by Navy Wave. USN Photo (Lent).

A Seabee fife and drum corps at Camp Endicott. USN Photo (Lent)

News Shot from the South Pacific: Seabees assembling a pontoon causeway for landing their equipment. USN Photo (Lent)

A bulldozer helps haul ashore the equipment the Seabees will need to rebuild and defend a captured base in the South Pacific. USN Photo

Chow line on a South Pacific island during the rainy season. USN Photo (SEABEE/Lent)

Seabees man anti-aircraft gun as part of their training exercise. USN Photo (SEABEE/Lent)

Hand-to-hand combat training with bayonets. USN Photo

Obstacle #12 on the commando course at Camp Endicott. USN Photo (SEABEE/Lent)

Week-end liberty. USN Photo (SEABEE/Lent)

The Seabee on the left has just strung a telephone field unit under actual battle conditions. USN Photo (SEABEE/Lent)

Camouflage and concealment problem. Many men are concealed in this photo. USN Photo (Lent)

A 15" tree is swept away by the blade of a Seabee bulldozer. USN Photo (SEABEE/Lent)

The controls of a bulldozer keep both hands and both feet involved almost constantly. USN Photo

Caterpillar's steel treads dig into a steep up-grade. There isn't much that can stop a bulldozer. USN Photo (SEABEE/Lent)

Seabee bulldozer in action in the South Pacific. USN Photo (Lent)

This Seabee pontoon barge approaching a dry dock was assembled with individual pontoon compartments. USN Photo

This boulder is just another rock for this 18-ton Cat on the Seabees Proving Grounds. USN Photo

Seabees assembling a pontoon bridge. USN Photo (Lent)

"Keep the hook moving!" Seabee stevedores loading thousands of crates of oranges before going overseas. USN Photo (Lent)

Filling bomb craters in the runway come first on this bulldozer's work schedule. USN Photo (Lent)

Seabees in the South Pacific coordinate with Marine tank operator to blast holes for setting of dynamite charges. USN Photo

In a matter of minutes, the pontoon causeway is strung from ship to shore. USN Photo (Lent)

Completing a runway just prior to arrival of war planes. USN Photo (SEABEE/Lent)

PEARL HARBOR

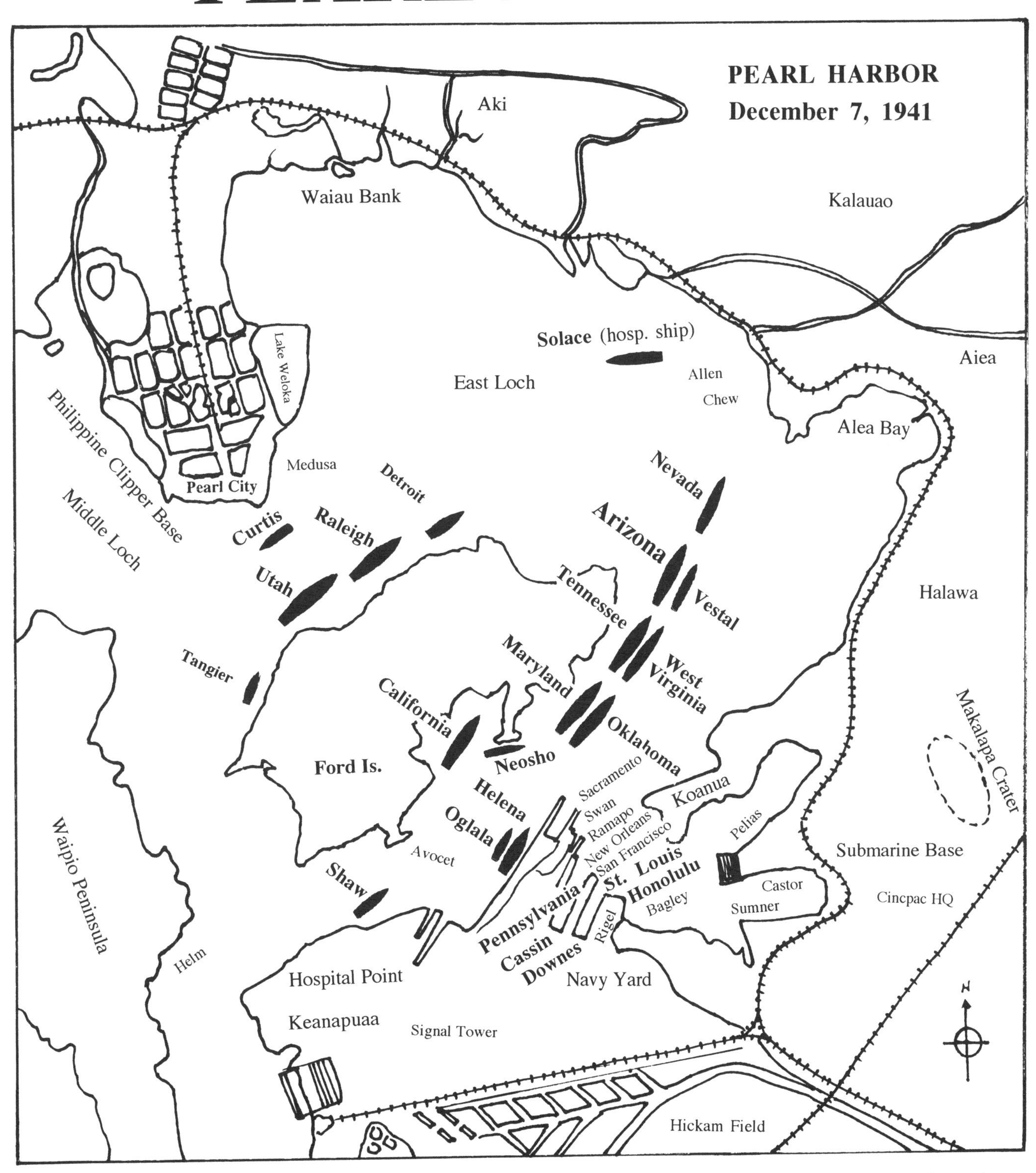

RESULTS OF THE ATTACK:

Battleships: *Arizona* (total loss), *Oklahoma* (total loss), *California* and *West Virginia* (sunk at their berths but later raised and repaired), *Nevada* (beached while under way to avoid sinking in deep water and later repaired), *Pennsylvania, Maryland* and *Tennessee* (all sustained lesser damage and were repaired). *Utah,* a former battleship (then a target ship), was sunk. Smaller ships: The Cruisers: *Helena, Honolulu* and *Raleigh* were damaged and later repaired. Two Destroyers: *Cassin* and *Downes* were damaged beyond repair. A total of eighteen ships were seriously damaged or sunk out of approximately ninety-six vessels in Pearl Harbor at the time of the attack. *Missing,* but not forgotten, was the battleship *Colorado* at the Bremerton, Washington Navy Yard and three indispensable carriers: the *Enterprise* was on its way to Pearl Harbor from Wake Island, the *Lexington* was ferrying aircraft to Midway, and the *Saratoga* was being repaired on the U.S. West Coast.

KOREAN WAR

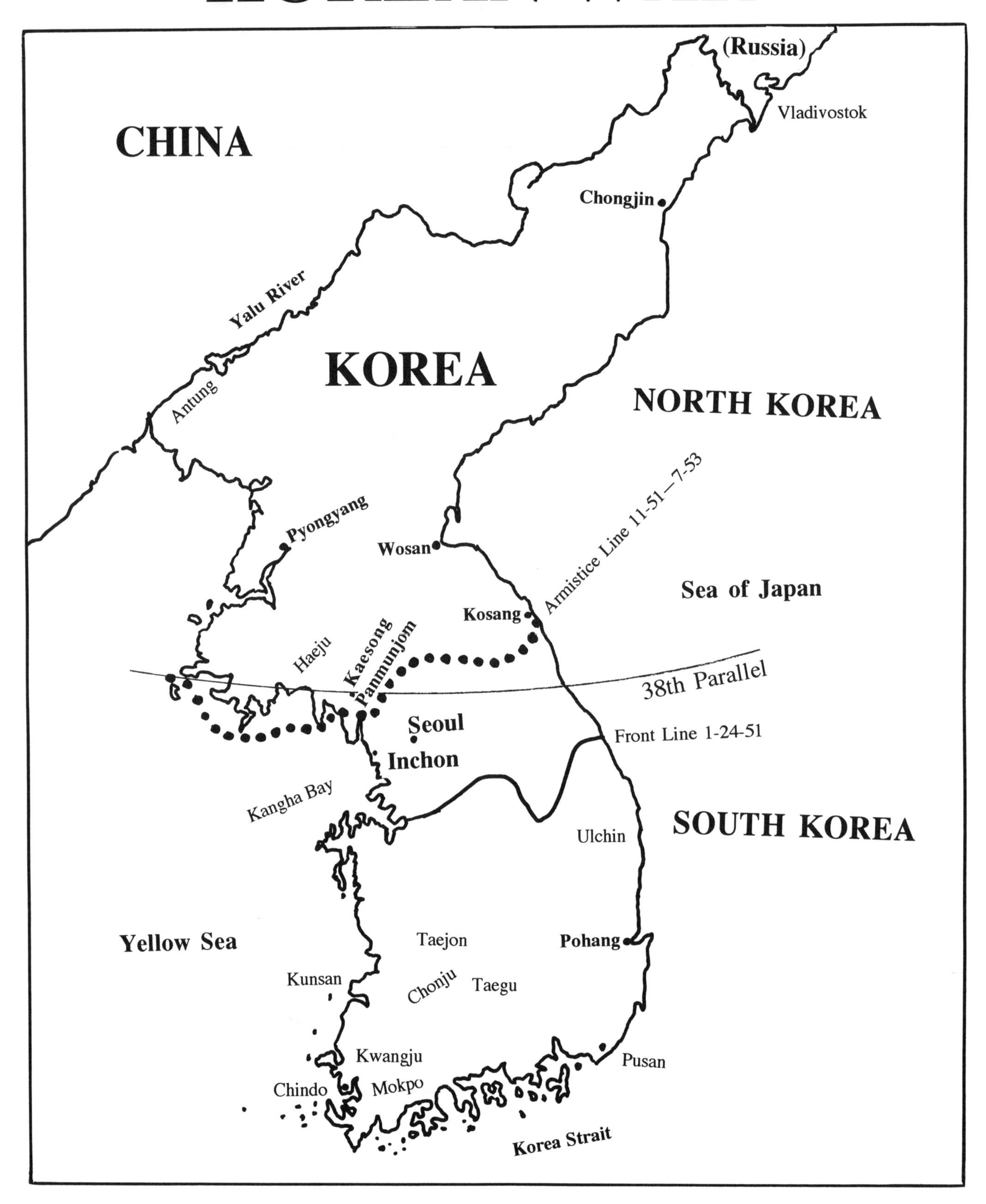

(Russia)
Vladivostok
CHINA
Chongjin
Yalu River
KOREA
Antung
NORTH KOREA
Armistice Line 11-51—7-53
Pyongyang
Wosan
Sea of Japan
Kosang
Haeju
Kaesong
Panmunjom
38th Parallel
Seoul
Front Line 1-24-51
Inchon
Kangha Bay
Ulchin
SOUTH KOREA
Yellow Sea
Taejon
Pohang
Kunsan
Chonju
Taegu
Kwangju
Pusan
Chindo
Mokpo
Korea Strait

VIETNAM WAR

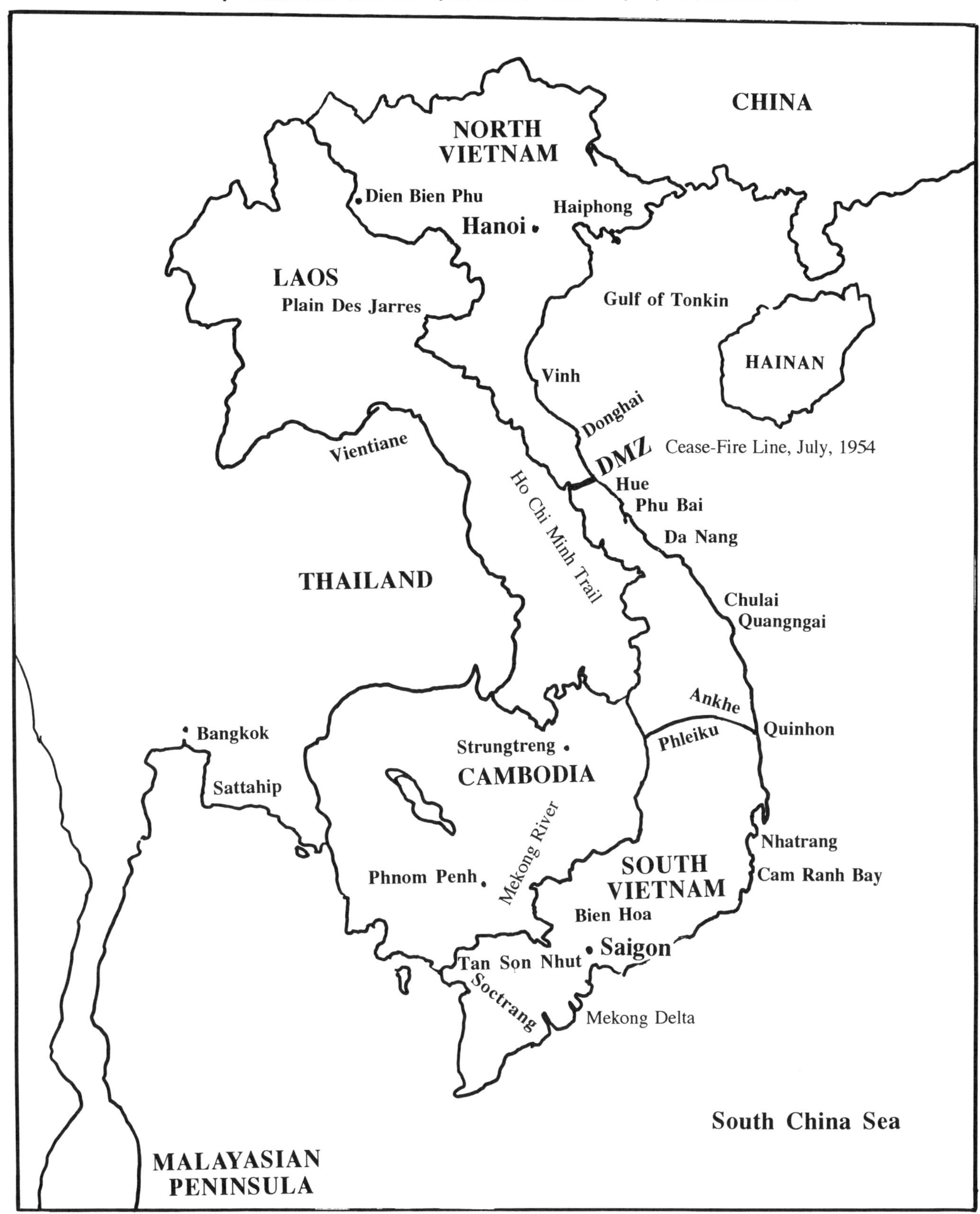
CHINA
NORTH VIETNAM
Dien Bien Phu
Haiphong
Hanoi
LAOS
Plain Des Jarres
Gulf of Tonkin
HAINAN
Vinh
Donghai
DMZ
Cease-Fire Line, July, 1954
Vientiane
Hue
Phu Bai
Ho Chi Minh Trail
Da Nang
THAILAND
Chulai
Quangngai
Ankhe
Quinhon
Bangkok
Phleiku
Strungtreng
CAMBODIA
Sattahip
Mekong River
Nhatrang
Phnom Penh
SOUTH VIETNAM
Cam Ranh Bay
Bien Hoa
Saigon
Tan Sọn Nhut
Soctrang
Mekong Delta
South China Sea
MALAYASIAN PENINSULA

DESERT STORM

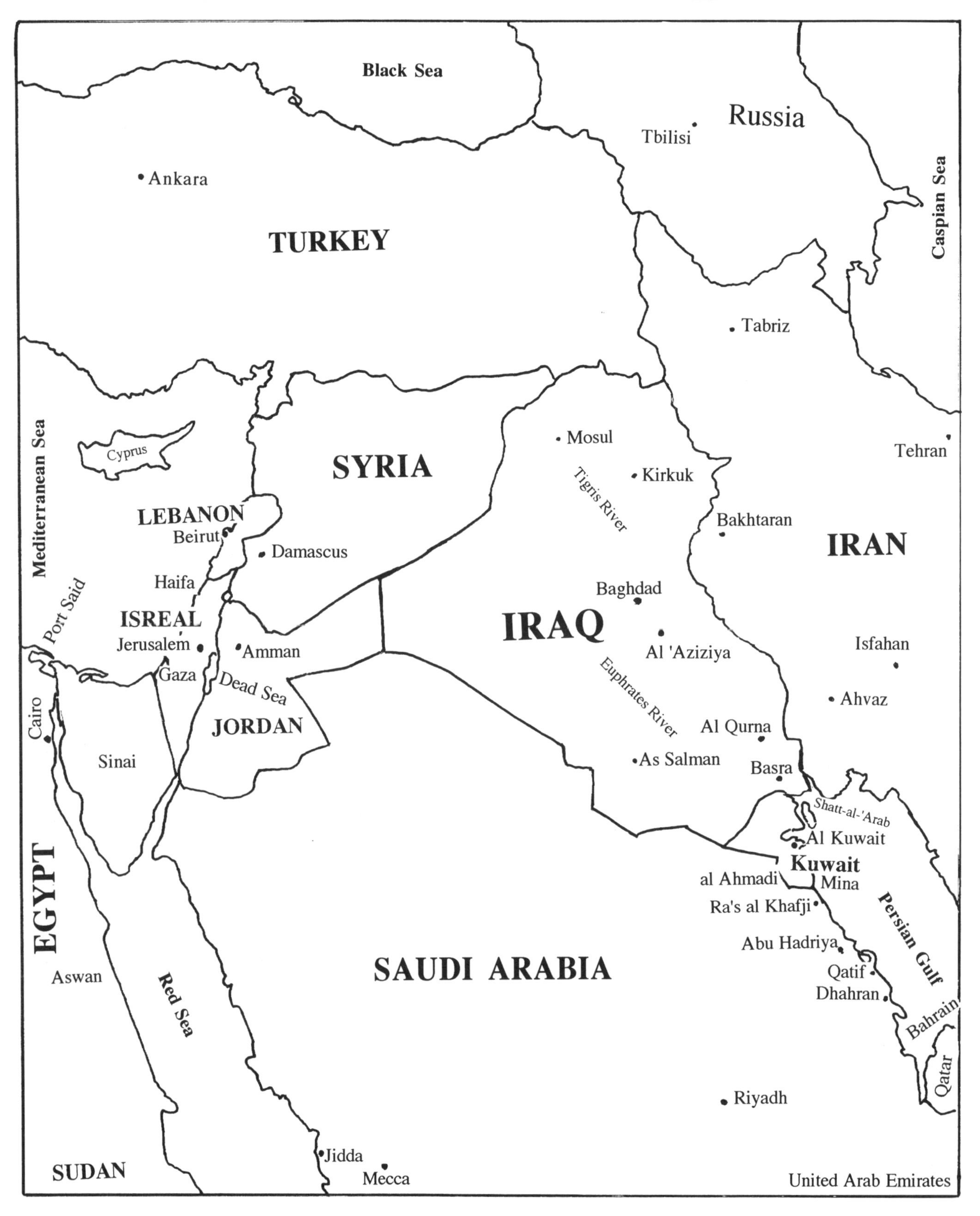

VIETNAM ERA SEABEE

Louis Schroen

BOOT CAMP

I'm reasonably sure the Seabees originated with World War II. My own experience began after boot camp with assignment to MCB 40. Like so many military abbreviations, that shorthand expression stood for mobile construction battalion. Many of the men went to the Seabee boot camp located at Gulfport, Mississippi, for five to six weeks of intensive training. It was during that time that they choose to join a particular battalion. Some men were promoted to instant petty officers because they had prior service or they had specialized job training for at least one to three years. They were given rates like E-4, E-5, E-6 or E-7 and the proportionate pay. We used the Navy ranks according to how many years of experience you had in job training.

WEAPONS AND OTHER TRAINING

Personally, after being sent to boot camp MCB 40 (a station in Rhode Island), I had to go through one month of intensive training on all available weapons. I had to qualify with M-16 and M-14 rifles before being deployed overseas. All of our training was closely supervised a Marine gunnery sergeant and a Marine captain. The training was pretty intense. You had to be there, or else face the wrath of the Marine drill instructor. We could not miss any meetings or inspections by Marines, period. We had some pretty intense sessions. Some guys couldn't hack it and had to be reassigned or discharged.

Sometimes, the personnel inspections lasted for eight solid hours, and you just stood at attention for eight hours on a "grinder"—a cement slab. We were ordered to stand at attention in the direct sunlight until we were dismissed, or passed out. Actually, at one time we had fifteen to twenty people pass out from the heat and the immobility. We had medics standing by to pick you up and carry you to the hospital. After that experience we began field training. We were moved up into the mountains and we stayed overnight in the snow-covered woods because all our training was completed during the winter. We simply stayed in the snow with very little protection from fox holes and often wet sleeping bags. Our clothing and gear got wet and remained that way. Unless you were extremely active, you were extremely cold . After test training, some people were chosen to go to Camp Le Jeune for more training on the M-60 machine gun and other weapons. Following that experience, the newly trained Seabees who were about to be deployed out of the United States were given an opportunity to get in a little rest first.

FLIGHT TO VIETNAM

We were transported to Viet Nam aboard U.S. Air Force jet cargo planes. Initially, we flew nonstop from Rhode Island to Washington State. On the way to Washington, one guy cut his wrists and had to be immediately treated by medics on board before he bled to death. It's doubtful that he was ever sent on to Vietnam. They took him off the plane the moment we landed and that was the last I ever heard about the guy. Shortly after we landed, another guy ran in front of a vehicle and was killed. It didn't appear to be an accident and it did appear that he was willing to risk his life to avoid going to Vietnam. We left the state of Washington and flew directly to Alaska. On the way to Alaska we had major problems with the plane. One engine burned out and before we set down on the runway, we had other difficulties. At least once, all of us hit the roof and a light flashed on saying, "fasten your seat belts." We continued to hit turbulent air pockets that day. The pilot tried to get above the turbulence but just couldn't do it. Next, just before the runway, we had more difficulties. Our wheels wouldn't come down on the plane and one engine was completely disabled. The pilot said the plane was low on fuel and that he might have trouble staying on the runway because of the severe weather. A crew member came back and manually cranked the wheels down. The pilot couldn't fire up the disabled engine so he flew onto the airport with only three engines. We hit the runway okay, but it was kind of scary, because we slid sideways. It was a big cargo jet with almost no windows. Every second during the slide felt like it might be our last. When we stopped sliding, we ended up about fifty feet from a big bank of snow that had been piled at least forty feet high. We gratefully got out of the plane and we stayed at the civilian hanger overnight. It was really cold inside the hanger, perhaps 40 degrees or less. After they worked on the jet for many hours, we flew to Japan where they replaced the engine completely. From Japan we flew to Viet Nam.

VIETNAM

I think Saigon was where we first landed. It was dark and very confusing. After leaving that airfield we got on a C-130 that had no windows that I can recall. It was during the middle of the night and there was a frenzy of activity on the ground. In a smaller, propeller-driven plane, we were flown to an airfield at Da Nang. The plane's fuselage was an empty cavern. No seats. No nothing. You just held onto a strap on the floor. The straps, obviously, were intended strictly for cargo. It was a very beat-up looking cargo plane. There were approximately one hundred of us aboard. I'm not positive. It was difficult to make an accurate count in the poor light. We then flew without incident into the advance base at Da Nang. They were bombing the runway as we arrived and we were informed that this plane would touch down but not stop. The pilot said he would taxi as slow as the plane would allow. A big, cargo door in the back was opened and an officer said, "You guys can jump off into one of the bunkers since we got other guys coming on board as soon as you hop off." The plane landed with a jolt and eventually went down the runway about five or six miles per hour. That was probably the slowest it could go. We jumped down and the big cargo door in the back shut. The plane quickly turned and took off down the runway.

SOUTH OF THE DMZ

There were no lights on because we were under attack and we just ran for any bunker we could find anywhere, any kind of protection, off the runway because the departing plane was dodging holes inside the runway lanes. In about thirty minutes the sirens for all clear blew and we reassembled our formation again. We got in "the camouflaged" six-bys (personnel carriers) and went north to a place called Phu Bai, about twenty to thirty miles from the DMZ At that time we were going north and had not been issued our rifles yet. We had no rifles, no ammo, no defensive weapons of any kind in case we were assaulted by the enemy. We had no way to protect ourselves, except for the ludicrous thought of hand to hand combat against concealed and heavily armed Viet Cong.

We were then stationed in our Seabee barracks. We called them hooches as they were really cramped and held about 12 people per hut. The night we got there, we heard that our Seabee camp was called Phu Bai. Loosely translated it meant "Valley of the Dead." It wasn't much before that, maybe a couple of weeks, that this same Seabee camp had been overrun by the Viet Cong. The Marines and Army had just taken it back so we were all really jittery at first. The next day we got things straightened out a bit. We were assigned to different job sites. Speaking for myself, I was assigned to the galley for about one month as a cook's helper. Later, I was stationed in my designated job as an ironworker. One of our first tasks was to make a security fence around what was to be a small hospital building. We then focused on building the field hospital itself. The builders in our company were divided into steelworkers and woodworkers. The first priority had been to complete a mile-long fence around the proposed building site for protection. Outside the surrounded area were Army and Ranger units and additional bunkers extending several miles in every direction. My job was to assist in erecting the steel frame and sheeting for the hospital building. Then, after working all day, we made new bunkers to replace our inherited bunkers inside the camp. One night they told us to put no roofs on the bunkers. Well, I decided, "Whatever flies up comes down," so I put a steel roof on ours with runway mat. That night a chief's hooch (private quarters), across from our hooch and about fifty feet away, was hit dead center by a rocket . Some people were injured seriously. We heard desperate yelling for help. Our roof, with its steel matting, was like armour plating. It was full of junk the next day from the chief's hooch. If we hadn't had that steel mat above us, we would have been hit by shrapnel and other lethal debris.

That night I was on guard duty and was written up for running to my bunker and leaving my guard post because I was supposed to just stand there. But on the next day there was another rocket attack and we were told to seek safety first. I found out later that at least one person was hurt seriously. Just how many others were hurt I don't know for sure. From then on the chiefs and officers all wanted a concrete, two-feet thick, poured roof on the bunkers they were making because they had seen my runway mat steel roof and they realized it might be the only way to survive. They started making roofs for all of the bunkers.

During my tour in Vietnam, which was about nine months (the usual length of a tour of duty for a Seabee), I learned that some of the guys were already on their third tour in Vietnam. I had joined the Navy for four years and under the circumstances they could keep sending you back. If you were in the Army they sent you for one year and that was the limit you could be stationed in Vietnam without re-volunteering. During our nine months in Vietnam we were hit by approximately forty-five rocket attacks. Amazingly, no one was hurt as a result of those particular attacks, that I know of, except for various cuts and non-disabling injuries. Mostly this was due to having a little warning. We had about sixty seconds warning from the time the siren blew until the rockets hit. We quickly learned the obvious lesson. We stopped whatever we were doing and moved fast whenever we heard the siren blow. You had to get into your bunker to have any real protection. It was not uncommon for buildings to receive a direct hit and be destroyed. The next day the old hassles over pilfering of supplies started up again. Eventually, an Army unit sent us an apology letter for the raids they had made on our stockpiles. The letter was read to us but didn't change anyone's behavior. Army and Marine units would steal our supplies constantly. We had to keep making new requisitions to the supply yard and at times just do without.

REST & RELAXATION

Other than the rocket attacks, we had it pretty safe. We went on two R & R's. Everyone got to go on an R & R for one week. Some men made arrangements for flights as far away as Japan. I just took my time off in Vietnam. We had two R & R times when the entire company went only for one day. On one occasion we drove all day to get to the Marine R & R camp. It was located four miles from the DMZ and was a heavily guarded area with beach frontage. We stayed there all day and enjoyed the sandy beach and warm, tropical waters. Mostly, we played in the water.

The distance from us, however, didn't justify taking all day to get there. It was only twenty-five miles away but you had to cross rivers and wait to take barges or landing crafts because there were no intact bridges.. The second time we flew and it only took about fifteen minutes in a helicopter. The R & R helped to relieve some of the stress of being in a lethal war zone. Thoughts of death and evidence of the death of others were all too common. Our camp itself was called Valley of the Dead. We lived in a graveyard. During rocket attacks there were many times when we couldn't find a bunker, or get to any bunker quick enough, so we just dived behind one of the big tombs that were scattered throughout the Vietnamese cemetery. These tombs were approximately thirty feet square and they had a one or two foot gap between the outer concrete wall and the inner concrete wall. It was easy to jump over these two to three feet high walls and crouch down during the rocket attack. The walls provided pretty good protection.

The Vietnamese bury their dead in a vertical position. Often we could see where they had buried someone in an upright position and then covered them with concrete. In our whole camp, we didn't realize it, but it smelled like a garbage dump all the time. It smelled bad—really bad. When we got there our water system was blown up so we had to make provisions for water. We had no showers for about two-three weeks, except for what we could do by dabbing a little cold water around. Eventually our electrician's unit got the hot water system going. But all our water over there, like most of the camps, was brown-colored and had bugs in it. We did have good food and most of the Marine and Army units who were in our camp got all they could eat. We had 800 men but we usually fed about 1,200 to 1,300 men every meal. You can see how some of their people came to our camp to eat because some of their food, especially that of the Marines, was not very good.

DEPARTURE

After nine months in Viet Nam we got out of there and were very grateful to still be alive. I'm pretty sure we got back home in time for Christmas. Our unit first made its way south to Saigon aboard the lumbering cargo planes, C-130's. We flew out of Saigon on a chartered, commercial jet. I can't recall the particular airline company. The plane was a standard, four-engine jet and actually had seats. The civilian pilot noticed people were shooting at the plane with ground fire. We could see the bullets flying as every fourth bullet was a tracer. He simply put the jet into a full take-off position and accelerated just as fast as it would go. Once in the air he climbed as steep as it would go until we got to at least 40,000 feet. They don't typically fly at that altitude in the United States at all. It was potentially real nose-bleed country if the pressurization failed, but we didn't care. Moments before, we thought it would be worse than ironic to be shot down on the day we were leaving Viet Nam. On the way up the whole jet was shaking like crazy because he apparently had the engines wide open. The plane seemed to be tilted as steep as it would go. We just held on and didn't blame him one bit. After we got off I don't think we were hit by gunfire anywhere on the jet because we made it back to Japan without further incident.

We soon got back to the states and had to start our intensive, Seabee training again after being stationed outside of the United States. The basic rule seemed to be that you had to start your same training over again before being deployed anywhere else. Our next deployment was supposed to be Diego Garcia (an island in the Indian Ocean) but it wasn't. Instead, we were sent to Puerto Rico. There, we did small jobs and remained there for nine months. We had too many people for the jobs that were required of us. We had 1,400 people for the barracks situation and obviously didn't have the continuous hazards of being in a hot war zone. The solution was to deploy some of us all over the world. Some went to Panama Canal, some worked for the U.S. president making tennis courts and swimming pools and "executive" things of that nature in the capitol. We worked for a telephone company and did various odd jobs in Puerto Rico for nine months. And then from there we went back to the United States, and completed the same training—again.

After five weeks of severe training we were then sent to the island of Diego Garcia. Some of the supplies had already been sent there because it took a couple of months by ship for supplies to arrive, such as the water purification plant which made salt water into fresh water. It didn't always work as well as it should have. On our way there we flew under the same arrangement as the trip to Viet Nam. From Rhode Island to Washington State, to Alaska, to Japan, to the Philippines, and maybe one other island. Then we got onto a ship for four days. At first the ship was serving us lousy, lousy food. People were getting sick all over because it was also very hot. It wasn't seasickness because the ocean wasn't rough at all. It was a flat-bottomed ship and rode like a sea-going barge. It was a big landing-craft type of ship with a big cargo hold. It could probably hold at least one hundred tanks. Our captain talked to their captain and we got good rations and ate before the crew was served. Actually, from then on, we had very acceptable food. After landing on the island we had the next day off because we were had just spent four days on a slow-moving ship and three-to-four days in the air flying. We had spent seven days getting there and we were pretty seasick, air-sick, or otherwise not in the best of shape and could not start working immediately. So they gave us one day off before we were assigned to our new jobs.

This island had been owned by the British, and perhaps it still was; we had no idea. There was nothing on the island at all, except a few birds. About fifty guys got there before us. Those people slept in tents when they first got there. There was no way to take showers. We had to bathe in the ocean and use saltwater soap. With the strong current, uneven beach and constant action of the waves, it was typical to keep losing the soap. We had to bath that way ourselves for about one month until we got the water hooked up and could take fresh-water, outside showers. We had a long pipe set up that emptied into little, round swimming pools, like the plastic ones you see children using when they're playing out in their yards on a summer day. You took your shower right along the side of the road where the system had been set up.

It was a very busy island because we had established a runway there. That was our main objective in the first place. Three different shifts were set up to keep the equipment in use constantly, especially to keep the diesel cats running constantly. The area to be cleared had many coconut palm trees that were about forty feet high and had to be pushed over by the diesel cats. The gravel needed was obtained from the ocean. At low tide we would go out and drill holes and set charges. We would blow as many as fifty boxes of ninety-proof nitroglycerine dynamite at a time. The crane had rigging with a high-lead like they use in Oregon for logging. It was fitted with a big scoop and scooped the broken coral up and put it into Euclids (front-loaders). The coral was then transferred to a rock crusher. Actually, the gravel was one hundred percent coral. There was no solid rock on that island. Once pulverized the coral gravel was spread on the runway and packed down. I don't recall how many feet thick, but perhaps as much as three feet in some places. At the end you'd take a soil machine that takes ground up rocks and add cement and water and you get a cement surface about one foot thick. It took one year to build the runway. It was almost two miles long when we left. The last I heard, that same runway has now been extended to about four miles in length.

MAIL CALL

On the island letters were a little problem because they come by merchant marine ship. It took one month to get home and one month to get back at the soonest. The absolute minimum to send a letter and get an answer was two months—often longer. We had some radio communications. We could wait at night to get access to a radiotelephone but mostly you'd have to wait night after night, after night. Your turn would eventually come up and then you might not get through and lose your position in line. We could say no names on the air, no locations, and little description of what anyone was doing or the persons monitoring every call would simply cut us off. We could say, "hi," to our wives and kids, or whoever it was we were calling, but no profanity of any kind. Any rough language and we were immediately cut off.

My job on the island was basically making small, tin buildings. Of course, some of them were pretty big, such as the mechanic's building. It was about 800 feet long and was to be used as a repair shop for virtually everything on the island. Both sides were open. Mostly it was framed and roofed in a way to keep the intense sun from making it almost impossible to work on metal objects. It did have a cement floor to facilitate moving equipment in need of repair in and out, and to be able to work underneath the equipment. The weather was sometimes reported at 130 degrees, but we later found out it was only 125 degrees—it just felt like more. Of the total time I was on the island (exactly one year) it only rained for two days. The official record keepers reported that it was 85 degrees on those two days. The rest of the time it climbed right back to the 130-degree maximum each day and averaged about 100 degrees every night. Except for the two memorable days of rain, no clouds were observed in the sky. It was crystal clear, as there was no evidence of pollution either. Every night you could see twenty to thirty shooting stars without much concentrated effort. Of all the different places where I've spent time around the world, there was no place where the sky remained as crystal clear, both day and night.

Naturally, for that very reason the island of Diego Garcia has the world's most visually accessible locale for both take-offs and landings. A small beach area was set up for us to go swimming with only a low risk of injury by sharks. There were white, sandy beaches and ideal-temperature waters. We always had to be careful in the water because of the potential for sharks. The most common sharks near the island were one of two types—white and tiger—the two most dangerous sharks for human swimmers. Despite the hazards, we did go snorkeling quite a bit on our few times off. We did have quite a bit of fun in the water and there were no incidents of serious injury that I can recall. While stationed in Puerto Rico many of us had acquired a great interest in snorkeling.

MILITARY DRESS

We did have a military clothing shortage while I was stationed on Diego Garcia. As a result, there was no uniform dress code at all. No inspections, no haircuts, no real desert or tropical clothing. Soon the guys made their own shorts out of a regular pair of pants, or blue jeans. We just wore sneakers and plain T-shirts on the job. Some of us had army shorts and they were worn constantly. We were a motley crew. As a group we looked like the cast on the television comedy called *F-Troop*. Our whole battalion would meet each morning. Even the officers, who still had sufficient military clothing, looked liked they'd slept in their clothing for the past two months. Their clothing appeared clean but there was simply no way to iron or press anything. The island's laundry unit was hard on clothing and that's simply the way they got their clothing back. I didn't even have combat boots at the time. Sneakers, however, were cooler and more comfortable than any high-top, leather boots could have been. Since there were no inspections nearly everyone went without a haircut for the entire year. Before they would allow us to return to the United States we had to get out scissors and trim each other's hair up a little. On that island we did have a drinking problem. They said the rationed supply of beer was sufficient for each man to drink the equivalent of a six-pack per night. It was probably true because there wasn't anything to do on the island. No entertainment; no nothing. We had movies at night (when the projector worked), but basically you worked at least an eight-hour day. You would go eat chow at each of the meal times, watch a movie for about two hours in the evening and then go to bed.

FLARE-UP

We had a club set up at first. It was just a big tent in the beginning. The purpose was to get together and do something a little different. Then we made a framed building for the same purpose. All of our food, when we first got there, was cooked with kerosene oil heaters. There was no complaint with the food . It was delicious every day and one of the few things you could really look forward to. One time we had a battalion party and we thought it was a good idea but we forgot we didn't have screens or covers and the flies were really bad. Out of 1,400 men at least 800 were sick with dysentery the next day from eating that exposed and contaminated food.

The toilets became the most prominent feature of the island and soon over-ran. It got worse and worse before it got better. The medical staff told us to mostly eat peanut butter and we were given some medicine. It was almost one week before it went away. All of us were leery about what we ate for a while.

One night they had something happen in the club and a chief shut down the club. Several guys became incensed as a result and drove the chief's Jeep into the ocean. It was major entertainment to observe. A Jeep will go quite a ways in the water. Maybe four feet into the water before the carburetor floods out and the engine stops. The chief got the OD and the OD came down to the beach. Soon he discovered the same thing had been done to his Jeep and they never did figure out who had done it. The OD went and got the captain and he came down with his Jeep. At that time he went to see what the problem was and he left his Jeep unattended also. He looked back and noticed his Jeep pulling away. The captain proceeded to get another Jeep and started after his original vehicle, but he was unable to catch the culprit who was, by then, cruising through the jungle. Eventually, this third Jeep was discovered buried in the ocean until it flooded out and stopped.

That night, true or not, I was told that our battalion commander called back to the United States and asked what to do. The stateside people (apparently) told him to get his battalion back under control or all officers would be relieved of their duties. There were approximately forty-five officers stationed on the island. The result was that the commander re-opened the club that night and everything went pretty smooth. There was also a burning of the chief's steps that night. Each building had about two or threes steps to get into it. The chief's steps had been removed first so the building wouldn't catch on fire. The incident put a touch of fear in all of the chiefs, or so it seemed, because there were so many enlisted men and it looked like the beginnings of a brawl. I can only speak for myself and I can only say I wasn't involved in it at all at the time.

Several fires had been started on the beach and the chief's steps had been added to the fire. The living quarters had tin roofs and made a lot of noise when guys threw branches up on the chief's roofs to keep them awake all night. It only happened for one day and night. Afterwards everything went along as if it had never happened.

I didn't know until recently, but a friend of mine, a former Marine, was over there recently and said the place where we worked on the island now is earmarked for NASA space shuttle re-entry, should there ever be a problem. Diego Garcia is located between India and East Africa, 100 miles below the equator. The weather is ninety-nine percent perfect. No rain, no clouds, no wind, just heat. The island itself is shaped like a horseshoe. It has a distinct bay area that is very well protected from the weather and ocean currents.

I didn't realize it until noticing in a current book *(It Doesn't Take a Hero)* that General Norman Schwarzkopf, Commander of U.S. forces during the Gulf War, noted that after the Viet Nam War it was his idea to put ships in the protected, Diego Garcia bay in case they were needed in that area again. Ships stationed there were, in fact, utilized during the Persian Gulf War.

That ended my three tours with the U.S. Navy Seabees. When I left the island of Diego Garcia, I had been in the military just over four years. I had signed up under the Hundred Day Program. Days turned into years. I got back into the United States in one piece and have continued working in the construction field. My Seabee experiences will never be forgotten. •

Louis Schroen [far right] with a couple of Seabee buddies. Photo taken in front of bunker and houch at Phu Bai, Vietnam, 1969 (probably during the month of March).

Observation Building at either St. Thomas or St. Troix, May, 1970. Building was designed with steel shutters to prevent explosive damage. Site was used as a radio control center to operate mobile tanks trying to elude Marine helicopter gunships (practice maneuvers).

Seabee built runway on the island of Diego Garcia. Pilings on either side.

Naval ship that transported much of the cargo and men to Diego Garcia, 1970.

Barges used to haul cargo off Merchant Marine and other ships, Oct., 1970.

Sea-going barge used to convert salt water into drinkable water.

VIETNAM ERA SEABEE

SEABEES IN ACTION

Seabees Constructing a "Sand" Roadway, Camp Peary, VA. "'Seabees'—those men whose job is to build and fight with the Navy—put together a roadway on the sand dunes of the Virginia coast where they are getting advanced training for work at overseas bases." USN Photo.

World War II Postcard—Construction Battalions Practice Landings, Camp Peary, VA. Seabees making landings from surf boats during their training course. USN Photo.

Courtesy of National Archives

SEABEES IN ACTION

Seabee Commando Trainees, Camp Peary, Va. "Seabees charge down a hill at Camp Peary, Va.where they are learning Commando tactics. These men work at building air bases, supply depots, and other other installations at advanced bases, and are always ready to fight." Official USN Photo.

Seabees Demolition Practice, Camp Peary, Va. "Seabees Dynamite a tree stump which obstructs a road building project at Camp Peary, Va."

Courtesy of National Archives

SEABEES IN ACTION

Seabee Postcard.

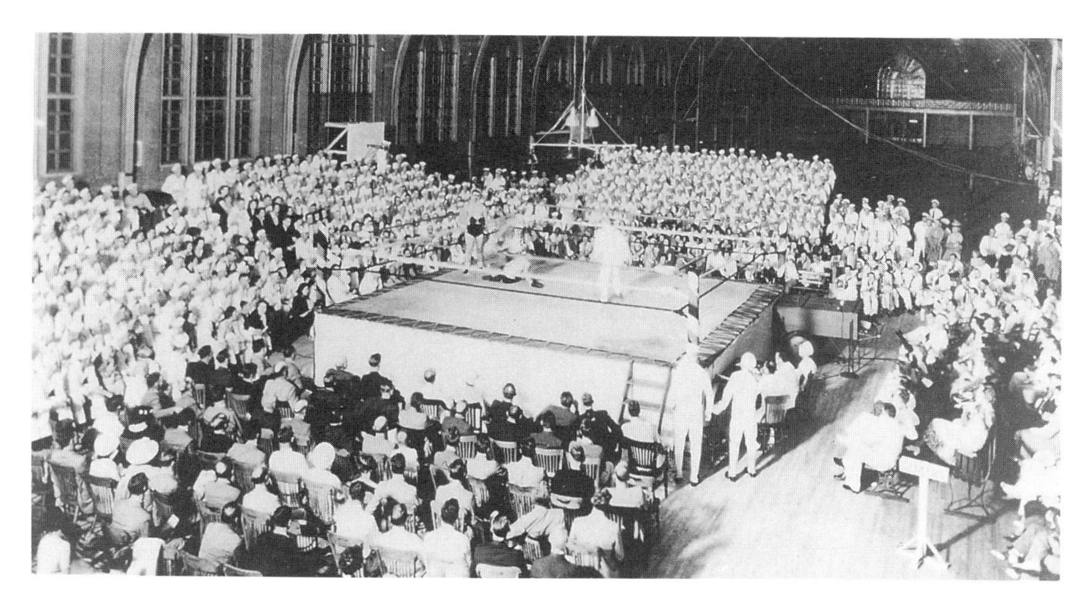

Happy Hour U.S. Naval Training Station, Great Lakes, IL. Rear Admiral John Downes U.S. Navy Commanding Officer. USN Photo.

Couresty of National Archives

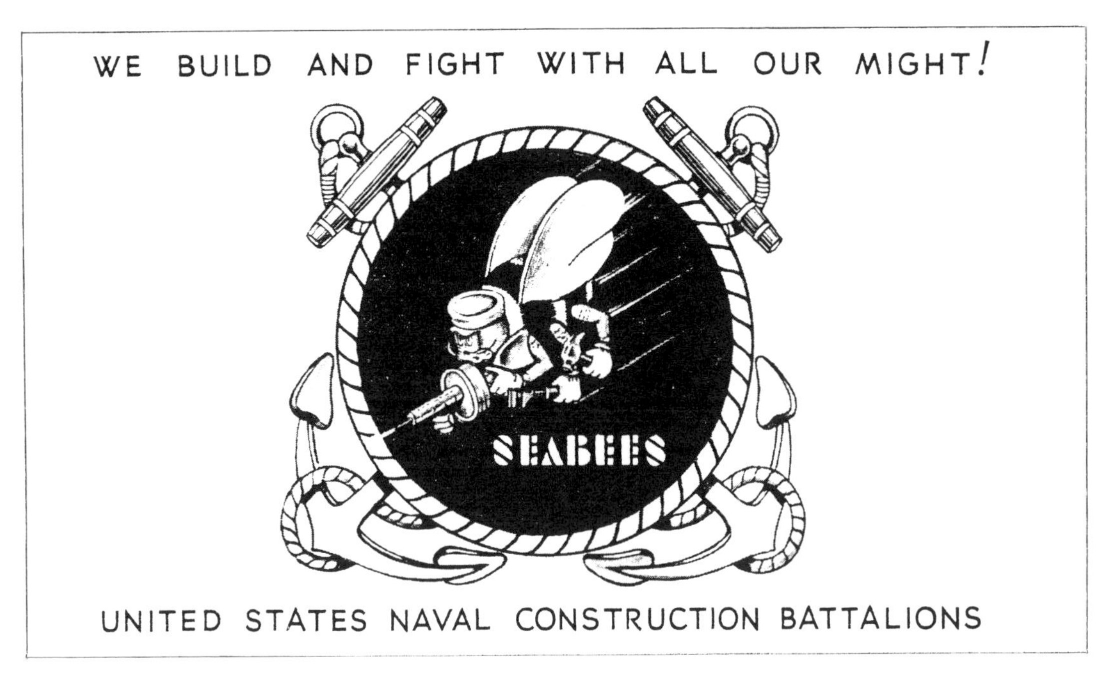

Seabee Postcard. Undated and postage free with following note: From: E. S. Lindberg, SK 2/C, 4th Reg. Repl. Group, Advanced Base Depot, Port Hueneme, Cal.: "Dear Son Charles: How is my big sailor son? I bet you would like to be with your Dad, so we could be in the Navy together, but maybe you better stay home and take care of grandma and Hazel. I know you will always be a good boy, when daddy gets back home we will have a lot of fun going places. I hope to be home when you start school and maybe I will be. Goodbye Son, Daddy."

The "Seabee" Center, Davisville, RI. The "Seabee" center currently provides home base facilities and logistic support for all mobile construction battalions operating in the Atlantic and Antarctic areas. They installed the first nuclear power plant in the Antarctic. Courtesy of Max Silverstein & Son, Providence, RI

Courtesy of National Archives

VICTORY IN EUROPE DAY
May 8, 1945

"All the News That's Fit to Print"

The New York Times.

LATE CITY EDITION

NEW YORK, TUESDAY, MAY 8, 1945. THREE CENTS

THE WAR IN EUROPE IS ENDED! SURRENDER IS UNCONDITIONAL; V-E WILL BE PROCLAIMED TODAY; OUR TROOPS ON OKINAWA GAIN

ISLAND-WIDE DRIVE

Marines Reach Village a Mile From Naha and Army Lines Advance

7 MORE SHIPS SUNK

Search Planes Again Hit Japan's Life Line—Kyushu Bombed

The Pulitzer Awards For 1944 Announced

GERMANY SURRENDERS: NEW YORKERS MASSED UNDER SYMBOL OF LIBERTY

GERMANS CAPITULATE ON AL

American, Russian and Fren Accept Surrender in Ei Headquarters, a Reim

REICH CHIEF OF STAFF ASK

Doenitz Orders All Military To Drop Arms—Troops in N —Churchill and Truman

By EDWARD KEN

VICTORY IN JAPAN DAY
September 2, 1945

"All the News That's Fit to Print"

The New York Times.

LATE CITY EDITION

NEW YORK, SUNDAY, SEPTEMBER 2, 1945.

JAPAN SURRENDERS TO ALLIES, SIGNS RIGID TERMS ON WARSHIP; TRUMAN SETS TODAY AS V-J DAY

HOLIDAY TRAFFIC NEAR 1941 LEVEL; 'GAS' IS PLENTIFUL

Exodus From City Is Greatest Since Pre-War Days but Congestion Is Avoided

GOOD WEATHER PROMISED

Times Sq. Takes V-J News Quietly

HAILS ERA OF PEACE

President Calls On U.S. to Stride On Toward a World of Good-Will

SALUTES HEROIC DEAD

Cautions Jubilant Nation Hard Jobs Ahead Need Same Zeal as War

By WILLIAM S. WHITE

PRESIDENT STRESSES LABOR DAY OF PEACE

JAPANESE FOREIGN MINISTER SIGNING SURRENDER ARTICLES

BYRNES FORESEES

Japan's Surrender Ordered

TOKYO AIDES

WAR COMES

Reprinted by permission of *The New York Times.*

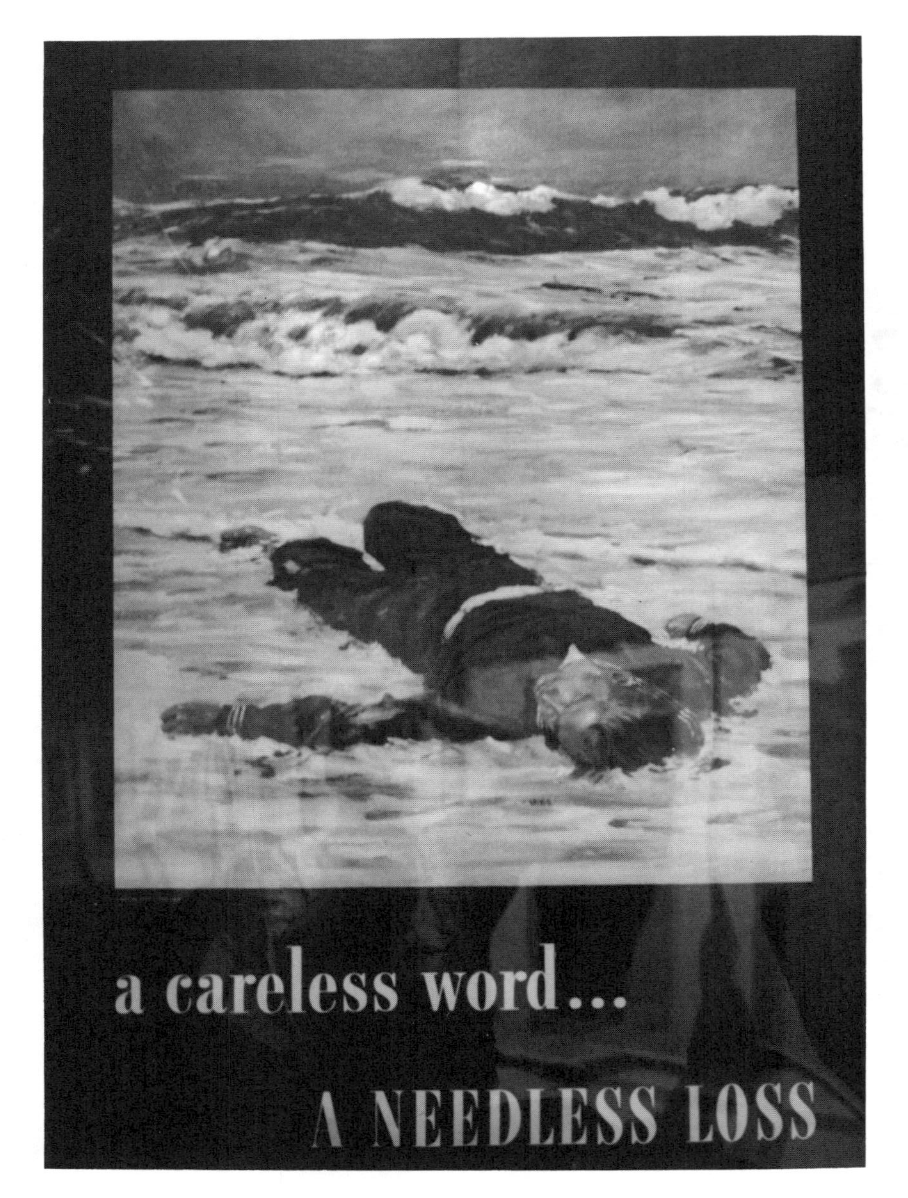

Courtesy of National Archives

Seabee pontoon dry dock

"Seabee Wave" recruited by the 20th Battalion. (See Dempsey cartoon).
USN Photo.

Magic (Pontoon) Boxes

by David A. Morrison, USN CPO PAD #1, #2 & #5 (Jan. 4, 1996)

Beginning in 1938, I was an electric welder at Puget Sound Navy Yard in Bremerton, Washington. I recall the date, December 7, 1941, vividly. I was listening to Bing Crosby and Rosemary Clooney that Sunday morning when the program stopped with the announcement that Japanese war planes were bombing Pearl Harbor and Hickam Field in Hawaii.

The next day, December 8, 1941, when I reported to work, there was a call put out for skilled workers to go to Pearl Harbor to help repair the damage. Three other single guys and myself volunteered. By January 1, 1942, we were on our way. I worked in a drydock and in a boiler shop on Ford Island until August, 1942, when I was sent back to PSNY in Bremerton. I continued to work there until September, 1942, and then I joined the U.S. Navy in Seattle, Washington. I was given a first class shipfitters rate in the newly formed Construction Battalions. I completed boot camp training with the 34th Seabee Battalion at Camp Allen and Camp Perry, Norfolk, Virginia. Later, about 100 men were sent by train to Gulfport, Mississippi, where the Navy Pontoon Assembly Detachment was formed as PAD #1. There, we taught the men to electric weld and to assemble the pontoons and to use the jigs that Chrysler Corporation had furnished [see David Morrison's article on page 124].

About the end of November, we returned to Norfolk, Virginia, where we boarded the *USS George Clymerand* and set sail for the Panama Canal. We spent Christmas Eve going through the Canal. On December 31, 1942, we all became Shellbacks after we crossed the Equator. On January 1, 1943, we arrived in Nomea, New Caledonia, and unloaded the ship at Ille Nou, New Caledonia This was an old French Garrison and the former Leprosy Colony. We proceeded to set up our Butler Building as our factory and set up camp using 8'by 8' Army tents as living quarters. We put in concrete floors and screening around the bottom. Before we went into production of pontoons, I was put in charge of beautifying the camp. Because of the primitive conditions and completely new environment, there was initially a lot of griping and complaining from the men.

We completed our first pontoon about July 1, 1943. It was during that summer that Lieutenant Reed formed a crew of men to go to Guadalcanal to assemble pontoons into drydocks and barges. Later that summer we dismantled our camp and shipped out to Bienka Island in the Russells. We set up camp near Sunlight Channel, built our pontoon factory out of 100' by 50' Quonset Hut material, and established our living quarters in 50' by 75' Quonset Huts with all four sides screened for ventilation. It became one of the finest camps in the South Pacific. Our men built our own steam kettles and ovens out of pontoons. We had showers in the heads and a mess hall that was large and clean.

PS. The CEC/SEABEE Museum at 1000 23rd Ave., Port Hueneme, California 93043, has a fine display of those Magic Boxes as they were soon called. Eventually, there may someday be pontoons on display at Gulfport, Mississippi, since that's where we started. D.M.

Seabees, Sailors, Soldiers and Marines search downed Japanese Dive Bomber for souvenirs. USN Photo.

Seabees of the 302nd Battalion operate pontoon barges off Tinian. USN Photo.

Seabee bridge construction, Vietnam, 1967. USN Photo.

Seabee operating a motor grader. USN Photo

Seabees of the 14th Battalion
Koli Point, Guadalcanal. USN Photo.

Seabees (94th Battalion) on Guam raising vegetables. USN Photo.

Seabees putting up a bridge on the airfield-beach road at Cape Gloucester. USMC Photo.

The Untold Story of the Pontoon Assembly Detachments • U.S. Navy 1942-45

by David A. Morrison, CPO, PAD #1, #2 & #5

With welding torches and 'jewelry,' Pontooner's lingo for hardware, men of the PADs stormed the beaches in the South Pacific to manufacture and assemble self-propelled barges, roadways of floating steel into docks, and piers for the invasion of the islands of Japan.

The steel pontoon was the brain-child of the Navy Yards and Docks. Admiral Ben Morrell and his engineering officers designed the first 5' x 5' x 7' steel, water-tight pontoon that could be linked together to make unsinkable barges, piers and drydocks. Chrysler Corporation manufactured the first ones to be used in Europe and Africa. The big problem was that they were bulky and took up valuable cargo space. The admiral and the officers of the Navy Civil Engineering Corps camp up with the idea, 'if Chrysler can build them, so can we.'

The Pontoon Assembly #1 was formed in Gulfport, Mississippi. In November, 1942, a complement of 450 officers and enlisted men took a month's advance training, and then shipped out to the South Pacific from Norfolk, Virginia, in December, 1942. They arrived in New Caledonia on January 1, 1943. PAD #1 then built their first camp and factory on the island of Ille Nou, New Caledonia. The T-6 pontoon went into production in the summer of 1943.

PAD #2 was formed on Bienka Island in the Russell Islands, 20 miles west of Guadalcanal in the Solomons. After setting up camp and the factory, the first pontoon, a T-6, was made effective January 31, 1944. PAD #3 was formed in Miline Bay, New Guinea, February 10, 1944.

PAD #4 reached Hollandia, New Guinea in November, 1942, and moved to Leyte in the Sumar area of the Philppines. On January 26, 1943, the camp and factory were established at Callicoan, Sumar.

PAD #3 shipped out from Port Hueneme, California. They left San Francisco and arrived in Los Negros, an island in the Admiralty Islands. A camp and a factory were built. Production was in full swing when Japan was bombed by the atomic bomb. The end of the war found PAD #5 on Guam, where it was inactivated.

When you consider the production records, PAD #1 built 1,152 T-6 pontoons in 1943, using 4,032 tons of steel. PAD #2, during a period of 22 months, from January 31, 1944 to November, 1945, built 4,424 T-6 and T-7 pontoons, using 15,750 tons of steel. Estimating what PAD #3, #4 and #5 produced, I would say the five PAD units produced 10,000 T-6 and T-7 pontoons, putting together 36,000 tons of steel. Besides building their camps and factories on those South Pacific islands, they assembled the pontoons into self-propelled barges, piers and drydocks. They even made ovens, water tanks and jails out of them!

The slogan, 'A NAVY SEABEE CAN-DO,' is typical of the many Seabees that served in World War Two in the United States, Europe, Africa, South America, South Pacific, and the Aleutian Islands.

In 1991, I learned from the Navy Bureau of Records, that the Pontoon Assembly Detachments had received a Navy Commendation, a Special Navy Commendation, and a Navy Citation from the Commander of the South Pacific.*

Seabees using empty gasoline drums for roofing. USN Photo.

The 12th and 23rd Battalions at Attu a few days after landing with the army. USN Photo.

AERIAL VIEW OF PEARL HARBOR

View of naval base at Pearl Harbor on October 30, 1941. The ships moored in the foreground are mostly destroyers. USN Photo / National Archives.

Seabee-built pontoon causeway at Normandy. USN Photo.

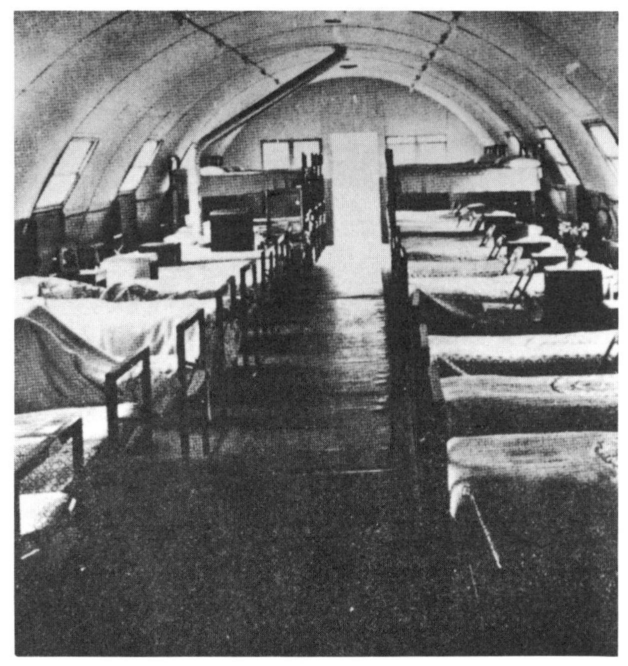

Seabee quonset housing, England, during WWII. USN Photo.

Nazi death cult leader.

Seabee Teams in Thailand. USN Photo.

Logo of 30th Naval Construction Regiment, Vietnam. USN Photo.

SEABEE COVERALL

HOME OF PACIFIC SEABEES

L. XXIV, No. 5 U.S. NAVAL CONSTRUCTION BATTALION CENTER, PORT HUENEME, CALIFORNIA 4 MARCH 196

Seabee publication, Port Hueneme, 4 March 1966.

Traditional Seabee logo patch.

Seabees attending a worship service under a canvas tent and while sitting on protective sandbags. USN Photo.

Seabees moving ashore with the first wave at Cape Gloucester, New Britain. USMC Photo.

Amphibious Seabee Base: Little Creek, VA USN Photo

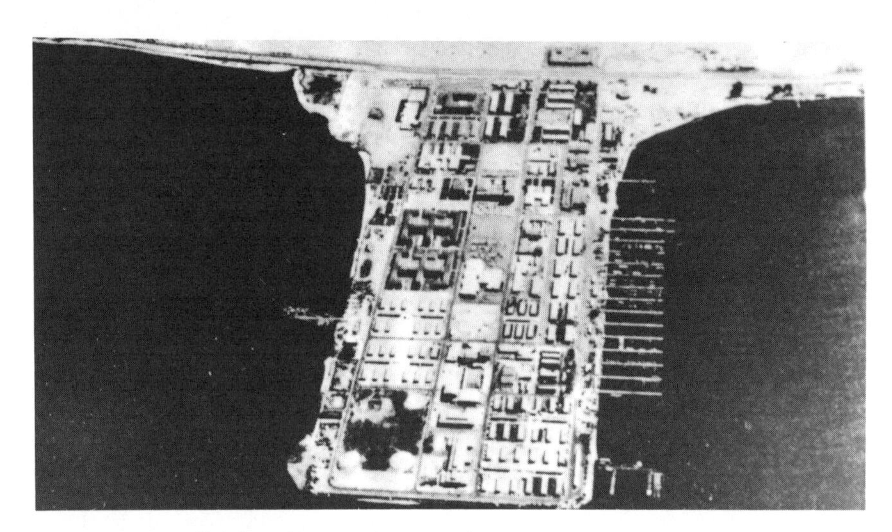

Seabee Base: Coronado, CA USN Photo

Seabee Base: Port Hueneme, CA USN Photo.

Pierheads off the Normandy beaches at Arromanches. USN Photo.

Men from the 7th Battalion assemble a pile driver for construction work in Okinawa. USN Photo.

Seabees constructing a weather station at Petropavlovsk, USSR, October, 1945. USN Photo.

Seabees present at Inchon Landing, Korea, September, 1950. USN Photo.

Seabee anti-aircraft crew (23rd Battalion) on watch in the Aleutians.
USN Photo.

Marine Pfc. Earl Brill of Pittsfield, Me., standing alongside (Seabee-built)
Marine Drive, Bougainville. USMC Photo.

Seabee operating a 40-ton roller on an airstrip in the Solomons. USN Photo.

Power shovel and trucks working from borrow pit. USN Photo

Admiral Nimitz signing off surrender papers aboard *USS Missouri,* at Tokyo Harbor, September 2, 1945. Also viewing (l. to r.) are Gen. Douglas MacArthur, Adm. William Halsey, Rear Adm. Forrest Sherman and a contingent of the ship's personnel. USN Photo.

Men of the 25th Battalion built the roads to Bougainville
USMC Photo.

Men of the 71st Battalion working about one mile behind front lines at Bougainville. Occassionally the work got ahead of the front lines.
USMC Photo.

Seabee-built naval air station receiving a carrier. USN Photo.

Pontoon experimental area at Davisville, RI, Seabee Base. USN Photo.

Iwo Jima Memorial, Wash., D.C. Sculpted by Felix de Weldon, former Seabee.

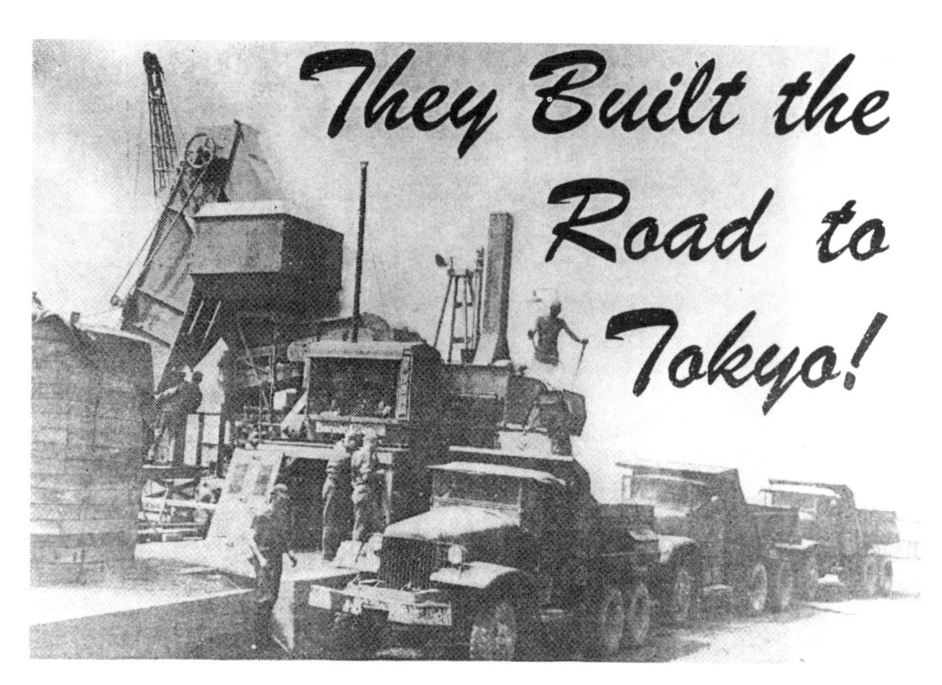

USN Photo.

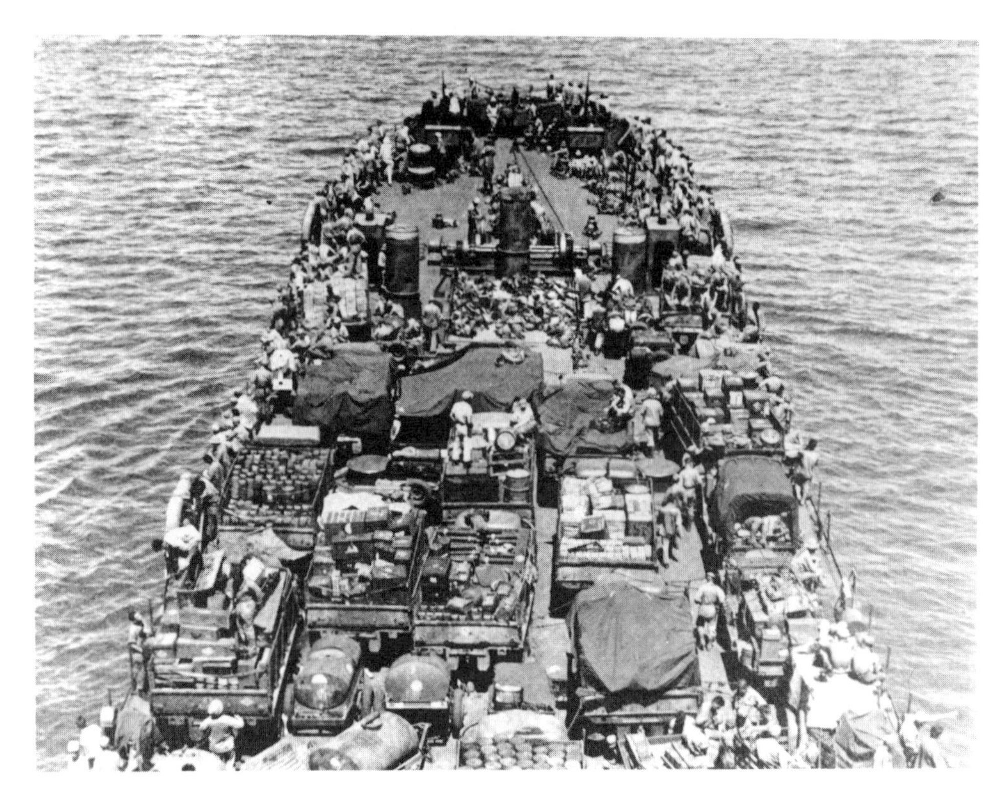

Invasion supplies. USCG Photo by Pho/M 1/c Don C. Hansen

Seabee Memorial. Dedication: "With Compassion for Others. We Build—We Fight. For Peace with Freedom. USN Photo.

CHIEF
Navy Bureau of Yards and Docks

(1942 through 1965)

NAVFAC COMMANDER

(1966 to present)

VADM Ben Moreell, CEC, USN	12-1-37—12-45
RADM John J. Manning, CEC, USN	12-1-45—12-1-49
RADM Joseph F. Jelley, CEC, USN	12-1-45—11-3-53
RADM John R. Perry, CEC, USN	11-3-53—9-25-55
RADM Robert H. Meade, CEC, USN	11-8-55—11-30-57
RADM Eugene J. Peltier, CEC, USN	12-2-57—1-30-62
RADM Peter Corradi, CEC, USN	2-12-62—10-31-65
RADM Alexander C. Husband, CEC, USN	11-1-65—8-29-69
RADM Walter M. Enger, CEC, USN	8-29-69—5-11-73
RADM Albert R. Marschall, CEC, USN	5-11-73—5-27-77
RADM Donald G. Iselin, CEC, USN	5-27-77—1-15-81
RADM William M. Zobel, CEC, USN	1-15-81—8-31-84
RADM John Paul Jones Jr., CEC, USN	8-31-84—8-14-87
RADM Benjamin F. Montoya, CEC, USN	8-14-87—10-27-89
RADM David E. Bottorff, CEC, USN	10-27-89—9-18-92
RADM Jack E. Buffington, CEC, USN	9-18-92—9-15-95
RADM David J. Nash, CEC, USN	9-15-95—Present

Admiral Ben Moreell (Seabee founder).

CURRENT SEABEE RECRUITMENT

Construction Force Naval Reserve

What Seabees Do

Navy Seabees train and work as carpenters, masonry workers, plumbers, steelworkers, equipment operators, mechanics, electricians and architects. The following jobs, called "ratings," make up the Naval Reserve Construction Force:

- **Builders (BU)** erect wooden, masonry and concrete structures and perform interior finishing work.
- **Construction electricians (CE)** install electrical and telephone networks; splice and lay wire and cables; and install, operate and repair generators, motors, transformers and lighting fixtures.
- **Construction mechanics (CM)** repair construction equipment and engines.
- **Engineering aids (EA)** draw plans, sketches and maps; survey land for construction; estimate material requirements; and test materials.
- **Equipment operators (EO)** operate heavy-duty construction equipment such as bulldozers and backloaders, and haul construction materials to building sites.
- **Steelworkers (SW)** weld and build steel and sheet-metal structures.
- **Utilitiesmen (UT)** install plumbing, purify water, operate boilers, repair ventilation systems and perform refrigeration and air-conditioning work.

In addition to these construction ratings, the Naval Reserve Construction Force also needs support-force men and women. These include boatswain's mates (BM), dental technicians (DT), disbursing clerks (DK), electronics technicians (ET), gunner's mates (GM), hospital corpsmen (HM), hull technicians (HT), journalists (JO), legalmen (LN), machinery repairmen (MR), masters-at-arms (MA), mess specialists (MS), operations specialists (OS), personnelmen (PN), photographer's mates (PH), postal clerks (PC), radiomen (RM), ship's servicemen (SH), storekeepers (SK) and yeomen (YN).

Service Requirements

Naval Reserve Seabees attend paid training periods with their units one weekend a month. One weekend a quarter, they may travel to their actual mobilization site, usually a major Navy base, for hands-on training in their construction specialties.

Once a year, reservists perform annual training (AT) for 12 to 17 days at a Navy installation. Reservists may be required to serve on extended active duty in times of national emergency or war.

Reservists may retire after 20 years of satisfactory service. However, retired reservists do not receive retirement pay and other benefits until they reach age 60.

The regular Navy formed its first Construction Battalion in 1942. Seabees served proudly in World War II, building roads, bridges, airport runways and even entire bases. Seabees have served in almost every world crisis and conflict since then. The first Naval Reserve Seabee units were organized in 1948.

The Seabees are known throughout the Navy for their "Can Do" spirit. Their construction feats, such as providing nuclear power to McMurdo Station in Antarctica, are legendary.

Navy Seabees are not only builders, but also fierce fighters. They can carve a Navy base from the wilderness, then defend it until reinforcements arrive to occupy it.

When mobilized, almost 16,000 Naval Reserve Seabees report to their units to help pave the way for the regular Navy.

SEABEES IN ACTION

South Vietnam—Navy Beachmaster's heavy forklift is off-loaded from utility landing craft 1481 as Beachmasters standby in waist-deep water. The forklift will be used in moving tons of Marine combat equipment over the beach in Operation Deckhouse VI. USN Photo by J. Means, USN (2-16-67).

An Hoa, Republic of Vietnam—Runway matting is unloaded from a U.S. Air Force C-130 Hercules Cargo Transport Aircraft for use by Seabees of U.S. NMCB-3 in repairing the airfield. USN Photo by PH2 R. C. Jones (12-1-68).

Port Mugu, CA—The MCB-5 Heavy Equipment Unit from Port Hueneme, CA, and the Oberg Construction Corporation work on the Calleguas Levy Break. The break caused flood damage at Port Mugu Naval Base. USN Photo by BV-2 Russ Bohnhoff (2-80).

Diego Garcia Is., Indian Ocean—A crane to be used in construction of a petroleum-oil-lubricants (POL) pier is driven up the sand ramp onto a barge. USN Photo (7-80).

SEABEES IN ACTION

Phu Bai Airport, Republic of Vietnam—Seabees of U.S. NMCB-133 resurface the runway as a C-130 Hercules Aircraft TA takes off from the completed end of the runway. USN Photo 7-68).

Chu Lai, Republic of Vietnam—Seabees of U.S. NMCB-58 work on the parking apron at the air freight terminal as an F-4 Phantom II Fighter Aircraft of Marine Fighter Attack Squadron 232 comes in for a landing. USN Photo by Lt. R. B. Luebke (4-12-69).

Norfolk, VA—Members of Amphibious Construction Battalion 2 lift sections of a causeway during exercise Elevated Causeway. USN Photo (6-86).

Norfolk, VA—A Marine, armed with an M-16 rifle, crouches on the beach in front of a utility landing craft. The LCU is transporting equipment to be used during exercise Elevated Causeway. USN Photo (June, 1986).

SEABEES IN ACTION

Camp Haskins, Da Nang, Republic of Vietnam—Seabees of U.S. NMCB-62 fill a revetment with sand at the Marine Aircraft Group-11 ramp. USN Photo by (unknown), No. K-86498 (July, 1970).

Binba Island, Republic of Vietnam—View of the shelters being constructed by Seabees of the U.S. NMCB Maintenance Unit 302 for Vietnamese Navy men and their dependents. USN Photo by W.B. Bass (8-2-70).

Norfolk, VA—A member of Amphibious Construction Battalion 2 directs the movement of equipment from a causeway onto the beach during exercise Elevated Causeway, a training exercise in which Seabees learn pier-construction techniques. USN Photo. SAVRIN: N1601 (6-86).

Norfolk, VA—A barge ties up to a pier constructed by members of Amphibious Construction Battalion 2 during exercise Elevated Causeway. USN Photo (June, 1986).

SEABEES IN ACTION

Guantanamo Bay, Cuba—A construction worker operates a grader at the site of the new Navy exchange mail complex. USN Photo by PH1 Gary Rice (2-23-87).

Charleston, SC—A Construction Battalion Unit 412 builder cuts wood to make a window frame for the Marine Detachment Club. USN Photo by PH2 Sean K. Doyle (4-87).

Norfolk, VA—A Caterpillar 988 Rough Terrain Container Handler prepares to tow a Helicopter Mine Countermeasures Squadron 16 (HM-16) RH-53D Sea Stallion helicopter. USN Photo (9-24-87).

Fort Hunter, Liggett, CA—Instructors prepare live fragmentation grenades for issue to Seabee's of NMCB-4 during the live grenade fire phase of the battalion's annual 20-day combat field exercise. USN Photo by J01 Phil Eggman (12-87).

SEABEES IN ACTION

Naval Station, Rota, Spain—A Seabee uses a circular saw to cut cross beams for a new section of flight line. USN Photo by PH1 Larry Franklin (5-89).

Tunisia, No. Africa—U.S. Navy Seabees shovel backfill over cement culverts as they repair a flood-damaged railroad during exercise Atlas Rail. USN Photo by PH3(AC) Stephen L. Batiz (2-90).

Tunisia, No. Africa—Equipment is unloaded from an Assault Craft Unit 2 utility landing craft which is positioned at the stern of the dock landing ship *USS Portland.* The equipment will be used during a Seabee construction exercise Atlas Rail, a Seabee project to repair a flood-damaged railroad. USN Photo by PH3(AC) Stephen L. Batiz (2-90).

Tunisia—A U.S. Navy Seabee uses an arc welder to solder metal bands on cement culverts during exercise Atlas Rail, a Seabee project to repair a flood-damaged railroad. USN Photo by PH3 (AC) Stephen L. Batiz (2-90).

SEABEES IN ACTION

Saudi Arabia—A CH-53E Super Stallion helicopter arrives over the Naval Mobile Construction Battalion 5 (NMCB-5) compound with a load of supplies. NMCB-5 is in northern Saudi Arabia to provide engineering support for coalition forces during Operation Desert Storm. USN Photo by CW02 Ed Bailey, USNR.

Saudi Arabia—A member of NMCB-5 uses a bulldozer to prepare area for camp construction during Operation Desert Storm. USN Photo by CW02 Ed Bailey, USNR, (2-91).

Saudi Arabia—A road grader used by members of a NMCB stands on a main supply route during Operation Desert Storm. USN Photo by J02 Pete Hatzakos (2-91).

Saudi Arabia—Waste cans taken from a latrine are burned at a forward camp being prepared by NMCB-5 prior to start of the ground phase of Operation Desert Storm. USN Photo by Ed Bailey, USNR (2-91).

SEABEES IN ACTION

Saudi Arabia—Seabees from NMCB-5 fill sandbags as they build a bunker near their camp's perimeter. USN Photo by Ed Bailey, USNR (2-91).

Saudi Arabia—A member of NMCB-5 stands watch as another Seabee operates a scraper. Seabees are preparing the area for camp construction during Operation Desert Storm. USN Photo by CW02 Ed Bailey (2-91).

Saudi Arabia—Members of NMCB-5 travel along a main supply route in an M-929 5-ton dump truck during Operation Desert Storm. The Seabee on the left mans a .50 caliber M-2 machine gun. USN Photo by CW02 Ed Bailey (2-91).

Saudi Arabia—Petty Officer 3rd Class K. M. Pahl Jr., a Seabee with NMCB-5 walks through a trench near the command bunker during a camp defense drill. USN Photo by CW02 Ed Bailey (2-91).

SEABEES IN ACTION

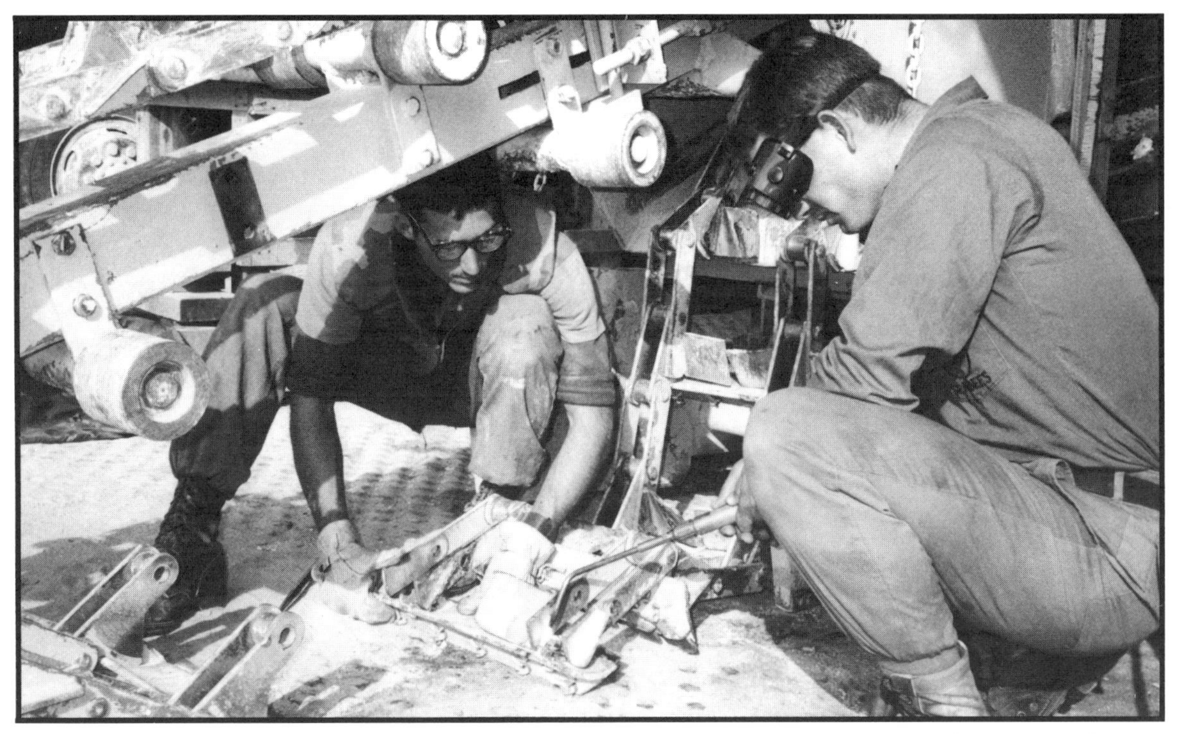

Saudi Arabia—A Seabee from NMCB-5 uses an acetylene torch as he helps to repair a trench digger at a camp under construction in northern Saudi Arabia during Operation Desert Storm. USN Photo by CW02 Ed Bailey (2-91).

Saudi Arabia—Mess specialists of a Naval Mobile Construction Battalion serve a meal during Operation Desert Storm. USN Photo by J02 Pete Hatzakos (2-91).

Saudi Arabia—A Seabee from NMCB-5 works on one structure as members of his unit finish the roof on another at a camp in northern Saudi Arabia during Operation Desert Storm. USN Photo by CW02 Ed Bailey (2-91).

Saudi Arabia—Members of NMCB-5 stand watch while another Seabee uses a bulldozer to prepare an area for camp construction during Operation Desert Storm. USN Photo by CW02 Ed Bailey (2-91).

SEABEES IN ACTION

Jubail, Saudi Arabia—A member of a U.S. Navy construction battalion backs up his M-49A2C fuel tanker truck at his unit's camp in the aftermath of Operation Desert Storm. USN Photo by J02 Pete Hatzakos (4-91).

Gulfport, MS—An equipment operator from a Navy Reserve Seabee unit dumps the bucket of a front loader into the bed of a dump truck during the Reserve Naval Construction Force "Readiness Rodeo" competition. USN Photo by JOSA Gary Boucher (5-91).

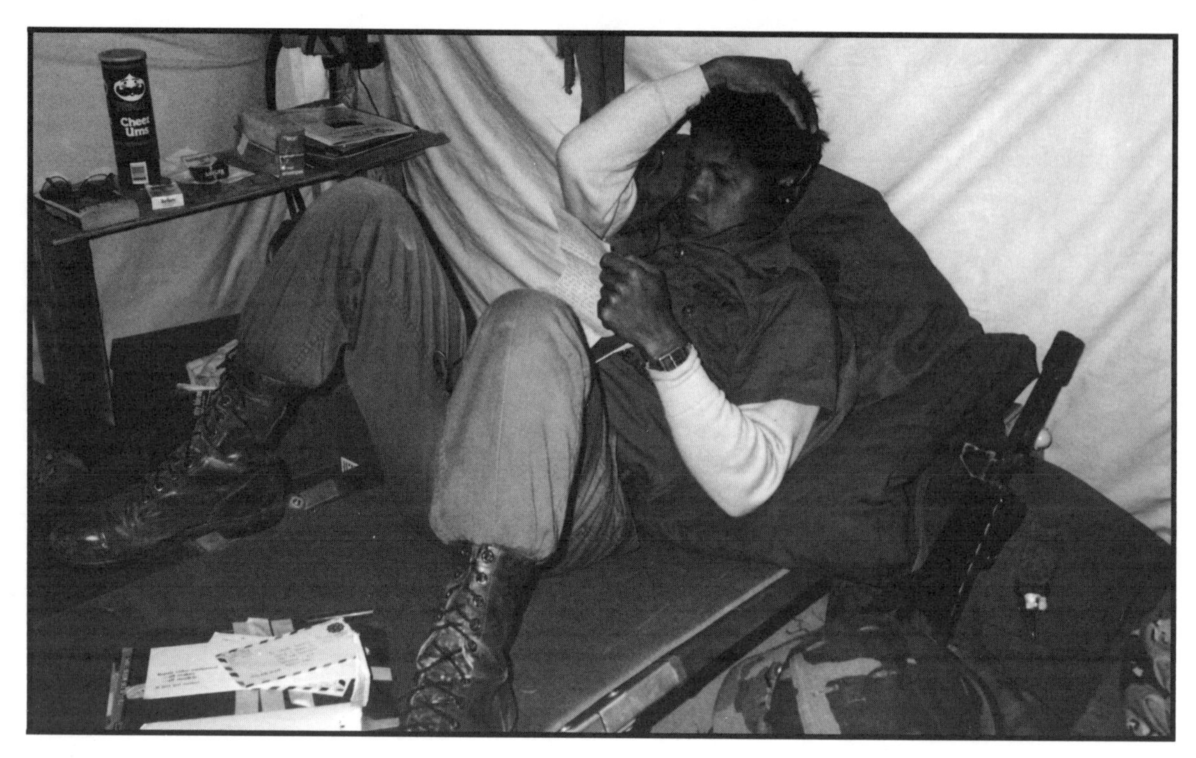

Saudi Arabia—Petty Officer 3rd Class Edgar Poblete, a Seabee with Naval Construction Battalion 5, reads his mail after work at a camp under construction in northern Saudi Arabia during Operation Desert Storm. USN Photo by CW02 Ed Bailey (2-91).

San Diego, CA—Storekeeper 3rd Class (SK3) Mian hands over new camouflage uniforms to Yeoman 1st Class (YNI) Williams. The new camouflage patterned uniforms are being issued to replace the solid jungle green work uniforms previously used by Seabees of Amphibious Construction Battalion 1. USN Photo by PH3 Brian McFadden (10-25-93)

SEABEES IN ACTION

Duc Pho, Republic of Vietnam—Seabees of U.S. NMCB-58 work on the airstrip as H-1 Iroquois Helicopters continue to lift off on combat missions. USN Photo by Lt. R. B. Luebke, April 12, 1969.

Norfolk, VA—Members of ACB-2 walk along a pier constructed by their battalion during exercise Elevated Causeway. USN Photo, June, 1986.

SEABEES IN ACTION

Norfolk, VA—Members of ACB-2 and their equipment move across a causeway during exercise Elevated Causeway. USN Photo, June, 1986.

Norfolk, VA—Steelworker Constructionman Grant Ramey, ACB-2, welds a pier piling during exercise Elevated Causeway. USN Photo, June, 1986.

SEABEES IN ACTION

Saudi Arabia—Timothy Lamm, a Seabee with NMCB-5, uses a grinding wheel in a workshop at a camp under construction in northern Saudi Arabia during Operation Desert Storm. USN Photo by CW02 Ed Bailey, USNR, Feb., 1991.

Rota, Spain—A Seabee prepares the ground for the foundation of a new building using a soil compactor. USN Photo by PH1 Larry Franklin, USNR, May, 1989.

SEABEES IN ACTION

Saudi Arabia—Petty Officer 2nd Class Aaron Marshall, NMCB-5, sets up a metal lathe in a machine shop during Gulf War. USN Photo by CW02 Ed Bailey, Feb., 1991.

Saudi Arabia—Petty Officer 1st Class "Doc" Jones, a Seabee with NMCB-5, takes a break while drilling a well at a camp under construction in northern Saudi Arabia during Operation Desert Storm. A 600-foot water well drilling system is in the background. USN Photo by CW02 Ed Bailey, USNR, Feb., 1991.

SEABEES IN ACTION

Saudi Arabia—A bulldozer and a scraper stand ready for use by members of NMCB-5 as they prepare an area for camp construction during Operation Desert Storm. USN Photo by CW02 Ed Bailey, USNR, 2-91.

Saudi Arabia—Tracy Holt, a Seabee with NMCB-5, stands guard at the entrance to the motor pool at a camp under construction in northern Saudi Arabia during Operation Desert Storm USN Photo by CW02 Ed Bailey, USNR, Feb., 1991.

SEABEES IN ACTION

Puerto Rico—Ensign Darius Banaji, resident officer in charge of construction, inspects a building wall at Naval Station, Roosevelt Roads. USN Photo by PH2 Suzette Lorene Hart, Sept. 7, 1988.

San Diego, CA—Seabees of ACB-1, Seaman Garcia and Steelworker Beamer check to ensure that the new camouflage uniforms for correct size and proper fit. USN Photo by PH3 Brian McFadden, 10-25-93.

NAVAL CONSTRUCTION BATTALIONS

1st Bat. Commissioned 3-15-42 at Camp Allen, VA. Served in South Pacific (New Hebrides). Inactivated 6-3-44.

2nd Bat. Commissioned 4-42 at Camp Allen, VA. Served in South Pacific (Funafuti, Tutuila, & Wallis Is.). Decommissioned 4-44.

3rd Bat. Commissioned 5-42 at Camp Allen, VA. Served in Pacific (Noumea, New Caledonia, Fiji, & Society Is.). Decommissioned 7-44.

4th Bat. Commissioned 5-42 at Camp Bradford, VA. Served in Aleutians Is., Guam & Okinawa). The 4th was in Okinawa at the time of the Japanese surrender, 8-45.

5th Bat. Commissioned 5-42 at Camp Allen, VA. Served in South Pacific (Midway, Palmyra, Christmas Is., Johnston Is, French Frigate Shoals, Canton, Kauai, Samar & Philippine Is.).

6th Bat. Activated 6-24-42 at Norfolk, VA. Served in South Pacific (Espiritu Santo, Guadalcanal, Tulagi, New Zealand, Noumea, New Caledonia & Okinawa.

7th Bat. Commissioned Spring, 1942. Served in South Pacific (Espiritu Santo, New Hebrides, Saipan & Okinawa).

8th Bat. Activated 5-23-42 at Dutch Harbor, AK then Camp Parks, CA. Served in South Pacific (Pearl Harbor, & Iwo Jima). Transferred to Hiroshima after war's end.

9th Bat. Commissioned 6-6-42 at Norfolk, VA. Served in Iceland, Pearl Harbor (Moanalua Ridge, NASD, Pearl City & Molokai). Transferred to Okinawa at the war's end.

10th Bat. Activated Summer, 1942 at Camp Allen, VA. Served in the South Pacific (Hawaii, Guam, & Samar).

11th Bat. Commissioned 6-42 at Camp Allen, VA. Served in Samoa, Noumea, New Caledonia, Ile Nou, New Zealand, Russell Is., Admiralties Is. & Subic Bay, Philippines.

12th Bat. Transferred from Port Hueneme, CA to Kodiak, AK, April, 1943. Also served in Dutch Harbor, Attu, & Adak. Inactivated 7-44.

13th Bat. Commissioned 7-13-43 at Norfolk, VA. Served in Dutch Harbor, AK, Akutan, Pearl Harbor, Tinian & Okinawa.

14th Bat. Commissioned 7-42 at Camp Allen, VA. Served in South Pacific (Noumea, New Caledonia, Guadalcanal, Espiritu Santo, Pearl Harbor & Okinawa).

15th Bat. Formed Summer, 1942 at Camp Allen, VA. Served in South Pacific (Espiritu Santo, New Zealand, Russell Is., Green Is., Pavuvu Is., Banika & Okinawa).

16th Bat. Commissioned 8-2-42 at Camp Allen, VA. Served in South Pacific (Funafuti, Ellice Is., Nanomea, Tarawa, Apemama and Makin in the Gilbert Is.

17th Bat. Commissioned 8-8-42 at Camp Allen, VA. Served in Newfoundland, Port Hueneme, Point Mugu, Saipan and Okinawa.

18th Bat. Commissioned 8-11-42 at Camp Allen, VA. Served in San Diego, Noumea, New Caledonia, Guadalcanal, New Zealand, Tarawa, Hilo, Saipan and Tinian.

19th Bat. Activated 9-42 at Norfolk, VA. Served in Noumea, New Caledonia, Australia, New Britain, Russell Is. and Okinawa.

20th Bat. Activated 10-42. Served in Noumea, New Caledonia, Woodlark Is., Oleana Bay, Vangunu Is., Viru Harbor, New Georgia, Kiriwana, Russell Is., Saipan and Okinawa.

21st Bat. Formed Summer, 1942 at Norfolk, VA. Served in Alaska at Dutch Harbor, Atka, Adak and Ogliaga. Also, Pearl Harbor, Moanalua, Intrepid Point, Waipio Point, Saipan and Ryukyus.

22nd Bat. Formed Summer, 1942. Served in Alaska at Sitka and Attu.

23rd Bat. Commissioned 9-4-42 at Camp Allen, VA. Served in Alaska at Cold Bay, Dutch Harbor, Atka, Adak, Kodiak, then Pearl Harbor and Guam.

24th Bat. Organized 9-4-42. Served in Noumea, New Caledonia, Guadalcanal, New Hebrides, Rendova, Kokurana, Baribuna, Munda, New Georgia, New Zealand, Russell Is. and Okinawa.

25th Bat. Activated 9-13-42 at Norfolk, VA. Served in New Zealand, Guadalcanal, Bougainville, and Guam.

26th Bat. Commissioned 9-18-42 at Camp Allen, VA. Served in Noumea, New Caledonia, Guadalcanal, Tulagi, Tupman, Calif., Kodiak, and Dutch Harbor, Alaska.

27th Bat. Commissioned Camp Allen, VA. Embarked to Tulagi on 1-3-43. Also served in Guadalcanal, New Zealand, Emirau, Bismarck Archipeligo, and Okinawa.

28th Bat. Commissioned Camp Allen, VA. First assigned to Iceland in December, 1942. Also served in Scotland, Le Havre,Calais, England, Okinawa and Yokosuka, Japan.

29th Bat. Commissioned 10-4-42 at Camp Allen, VA. Served in Rosneath, Scotland, England, France, San Clemente, Calif., Samar, Philippines and China.

30th Bat. Activated 10-42 at Norfolk, VA. Served in Trinidad, Dutch Guiana, Curacao, British Guiana, St. Lucia, Pearl Harbor, Samar and Philippines.

31st Bat. Activated 10-9-42 at Davisville, RI. Served in Bermuda, Hawaii, Iwo Jima (building roads to top on Mt. Suribachi) and Omura, Japan.

32nd Bat. Completed one tour of Alaska. Arrived Dutch Harbor 12-22-42, then Adak, Andrew Lagoon, and Camp Parks, VA.

33rd Bat. Served in Noumea, New Caledonia, Koli Point, Guadalcanal, Banika, Russell Is., New Zealand, Green Is., Russell Is., Palau Is., & Peleliu.

34th Bat. Commissioned 10-23-42 at Norfolk, VA. Served in Espiritu Santo, New Hebrides, Noumea, New Caledonia, Halavo, Florida Is., Guadalcanal, Russell Is., Tulagi, and Okinawa.

35th Bat. Commissioned 10-22-42 at Davisville, RI. Served in Noumea, New Caledonia, Espiritu Santo, New Hebrides, Russell Is., New Zealand, Lorengou, Manus Is., Camp Parks, and Philippines.

36th Bat. Formed in 1942 at Camp Allen, VA. Served in Espiritu Santo, New Hebrides, Banika, Russell Is., Noumea, New Caledonia, Saipan & Okinawa.

37th Bat. Commissioned 10-28-42 at Camp Endicott, RI. Served in Noumea, New Caledonia, Guadalcanal, Ondonga, New Georgia, Green Is. & Okinawa.

38th Bat. Formed 11-42 at Norfolk, VA. Served in Kodiak, Kiska & Adak, AK. Also at Elk Hills Naval Petroleum Reserve, CA, Tinian, Pearl Harbor plus Hiroshima, Kabayana, Yokosuka and Omura, Japan.

39th Bat. Commissioned 2-8-43 at Norfolk, VA. Served on Saipan until war's end.

40th Bat. Activated 11-42 at Davisville, RI. Served at Espiritu Santo, New Hebrides, Finschaven, New Guinea, Los Negros, Admiralties, Noumea, New Caledonia, Saipan and Okinawa.

41st Bat. Formed 11-30-42 at Camp Allen, VA. Departed for Kodiak, AK on 1-24-43. Returned to Camp Parks 3-44 then reassigned to Guam until war's end.

42nd Bat. Departed from Seattle 12-30-42 and arrived at Dutch Harbor, AK, on 1-5-43. Also served at Adak and Amchitka, AK, Camp Parks, CA, Pearl Harbor, Leyte Gulf and Samar.

43rd Bat. Organized 11-42 at Davisville, RI. Served at Kodiak, AK, Sand Point, Camp Parks, Oahu, and Maui.

44th Bat. Formed 12-1-42 at Norfolk, VA. Served at Espiritu Santo, New Hebrides, Manus Is., Noumea, New Caledonia and Okinawa.

45th Bat. Activated Fall, 1942. Served at Kodiak, Adak & Sitka, AK. Also, at Tonaga and Camp Parks.

46th Bat. Commissioned 11-18-42 at Camp Endicott, RI. Served at Guadalcanal, Finschaven, New Guinea, Los Negros Is., and Camp Parks.

47th Bat. Commissioned 12-7-42 at Camp Allen, VA. Served at Port Hueneme, & Bolinas, CA, Russell Is., Segi Point, New Georgia, Enogi Is., Munda, & Ondonga, New Georgia, Noumea, New Caledonia and Espiritu Santo.

48th Bat. Formed 11-42 at Norfolk, VA (Commissioned 12-15-42 at Camp Peary, VA). Served at Pearl Harbor, Maui, Puunene, Iroquois Pt., Oahu, Guam, and Rota, Marianas Is.

49th Bat. Commissioned 12-18-42 at Camp Allen, VA. Served at Bermuda, Davisville, RI, and Guam.

50th Bat. Commissioned 12-18-42 at Norfolk, Va. Served at Pearl Harbor, Midway, Oahu, Anguar, Palau Is., and Tinian.

51st Bat. Formed 12-2-42 at Davisville, RI. Served at Dutch Harbor, AK, Ulithi, Western Carolinas, and Saipan.

52nd Bat. Commissioned 12-6-42 at Davisville, RI. Served at Dutch Harbor, AK, Sand Bay, Great Sitkin Is., Adak, AK, Port Hueneme, Pearl Harbor and Guam.

53rd Bat. Activated 12-22-42 at Norfolk, VA. Served at Hadnot Point, NC, San Diego, CA, Camp Lejeune, NC, Camp Elliott and Camp Pendleton, CA, Noumea, New Caledonia, Guadalcanal and Guam.

54th Bat. Commissioned 12-24-42 at Camp Bradford. Served at Arzew, Algeria, Mostaganem, Cherchel, Port-Aux-Poules, Tenes, Benik-Saf and Nemours, Algeria. Also at Bizerte, Ferryville, Tunis, Karouba, La Goulette, La Perchie, Tunisia, Guiuan, Tubabas, Samar, and Cebu.

55th Bat. Effective 3-25-43 served at Brisbane, Australia, Merauke, Kanakopa, New Guinea, Port Moresby, Palm Is., Hollandia, Dutch New Guinea, and Mios Woendi Is.

56th Bat. Formed 12-24-42 at Norfolk, VA. Served at Pearl Harbor, Kaneohe, Oahu and Guam.

57th Bat. Commissioned 12-18-42 at Davisville, RI. Served at Espiritu Santo, New Hebrides, and Manus.

58th Bat. Arrived Vunda Point, Fiji on 5-4-43. Also served at Guadalcanal, Vella Lavella, Solomons, Auckland, New Zealand, Banika,Russell Is., Los Negros, Admiralties and Okinawa.

59th Bat. Commissioned 12-29-42 at Norfolk, VA. Served at Hilo, Kanuela and Pearl Harbor, Hawaii and Guam.

60th Bat. Formed 12-24-42 at Camp Allen, VA. Served at Brisbane, Australia, Townsville, Woodlark Is., Finschaven, New Guinea, Owl Is. Neocomfoor Is, Amsterdam Is. and Leyte, Philippines.

61st Bat. Formed 1-43 at Camp Peary, VA. Served at Espiritu Santo, New Hebrides, Guadalcanal, Auckland, New Zealand, Emirau, Bismarck Archipeligo, Russell Is., Leyte Gulf, Philippines, Guiuan and Samar.

62nd Bat. Formed 12-42 at Davisville, RI. Served at Pearl Harbor, Oahu, Maui, and Iwo Jima.

63rd Bat. Commissioned 2-43 at Camp Peary, VA. Served at Guadalcanal, Auckland, NZ, Emirau, Manus, Manila, and Philippines.

64th Bat. Commissioned 1-8-43 at Norfolk, VA. Served at Argentia, Newfoundland, Melbourne, Australia, Pearl Harbor, Samar, and Philippines.

65th Bat. Formed 3-43 at Freetown, Africa by merging CBD 1001 and 1002. Served at Boston, Mass. and Camp Endicott. Was inactivated 12-23-43 and personnel were transferred to other units.

66th Bat. Formed 1-43 at Davisville, RI. Served at Adak, Aleutians and Sand Bay. Second tour of duty at Okinawa and Nakagusuku, Japan.

67th Bat. Commissioned 5-18-43 at Camp Peary, VA. Served at Pearl Harbor, Tinian and Eniwetok, Marshall Is.

68th Bat. Formed 1-10-43 at Norfolk, VA. Served at Adak, Aleutians and Okinawa.

69th Bat. Commissioned 2-8-43 at Camp Peary, VA. Served at Argentia, Newfoundland, Plymouth, England, Falmouth, Dunkeswell, Omaha Beach, France, Vicarage, Southampton, Exeter, and Rosneath, Scotland.

70th Bat. Formed 5-27-43 at Davisville, RI. Served at Oran, No. Africa, Arzew, Algeria, Ain-el-Turck, Mostaganem, Tenes,Bizerte, Nemours, Beni-Saf, Port-aux- Poules, Salerno, Sicily Is., Pearl Harbor, Guam, Iwo Jima, and Okinawa.

71st Bat. Formed 1943 at Camp Peary, VA. Served at Guadalcanal, Bougainville, Manus and Pityilu, Admiralty Is., Los Negros and Okinawa.

72nd Bat. Formed 1-43 at Camp Peary, VA. Served at Pearl Harbor, Barbers Pt., Iroquois Pt., Ewa, Oahu and Guam.

73rd Bat. Departed from Camp Peary, VA, on 3-17-43 for Noumea, New Caledonia. Also served at Guadalcanal, Munda, New Georgia, Banika, Russell Is., Pavuvu, Peleliu and participated in the D-Day beach landings.

74th Bat. Formed 4-43 at Camp Peary, VA. Served at Pearl Harbor, Tarawa, Kwajalein and Okinawa.

75th Bat. Arrived at Noumea, New Caledonia during June, 1943. Later transferred to Guadalcanal, Bougainville, Banika, Milne Bay, New Guinea, Leyte Gulf, Philippines, Samar, and Calicoan Is.

76th Bat. Formed 1-43 at Norfolk, VA. Served at Pearl Harbor, Palmyra, Oahu, and Guam.

77th Bat. Commissioned 1-43 at Camp Peary, VA. Served at Guadalcanal, Vella Levella, Bougainville, Emirea, St. Matthais, Brisbane, Australia and Manila.

78th Bat. Formed 2-43 at Camp Peary, VA. Arrived Noumea, New Caledonia on 7-13-43. Also served at Milne Bay, New Guinea, Finschaven, Dreger Harbor, Los Negros, Admiralty Is., Manus and Okinawa.

79th Bat. Commissioned 2-1-43 at Norfolk, VA. Served at Kodiak, AK, Cold Bay, Amchitka, Adak, Saipan and Okinawa.

80th Bat. Formed 1-26-43 at Norfolk, VA. Served at Trinidad and Subic Bay, Philippines.

81st Bat. Commissioned 2-13-43 at Camp Peary, VA. Served at Rosneath, Scotland, Milford-Haven, Fowey, Penarth, Bicester, Falmouth, Salcombe, St. Mawes, Dartmouth, Newton, Abbot, Plymouth and London. Also at Utah Beach, Normandy, Paris, Pearl Harbor, Nakagusuku, Japan and Okinawa.

82nd Bat. Commissioned 1-28-43 at Camp Endicott, RI. Served at Guadalcanal, Vela Lavella, Munda, New Georgia, Ondonga, New Georgia, Sterling, Treasury Is., Nepoui, New Caledonia, Russell Is., Eniwetok, Ulithi and Okinawa.

83rd Bat. Formed 2-2-43 at Norfolk, VA. Served at Trinidad, Pearl Harbor, and Samar.

84th Bat. Commissioned 2-3-43 at Davisville, RI. Served at Brisbane, Milne Bay and Darwin, Australia, Biak, Morotai, Puerto Princesa, and Palawan Is.

85th Bat. Commissioned 2-6-43 at Camp Allen, VA. Served at Dutch Harbor and Attu, Alaska, Espiritu Santo, New Hebrides and Wake Is.

86th Bat. Formed 2-43 at Camp Allen, VA. Served at Adak, Great Sitkin Is., Amchitka, Tanaga, Andrews Lagoon and Okinawa.

87th Bat. Formed 2-23-43 Camp Peary, VA. Served at Banika, Russell Is., Treasury Is., Noumea, New Caledonia, Saipan and Okinawa.

88th Bat. Formed 2-8-43 at Camp Endicott, RI. Served at Mt. Dore, New Caledonia, Guadalcanal, Emirau Is., Ulithi, Leyte, Philippines, Samar, and Jinamoc.

89th Bat. Formed 2-43 at Camp Allen, VA. Served at Camp Peary and Camp Parks, VA.

90th Bat. Commissioned 7-25-43 at Camp Peary, VA. Served at Pearl Harbor, Angaur, Peleliu, Iwo Jima and Yokosuka, Japan.

91st Bat. Formed 10-21-43 at Camp Peary, VA. Served at Milne Bay, New Guinea, Ladaya, Hilimoi, Stringer Bay, Gili Gili, Madang, Palm Is., Australia, Finschaven, Brisbane, Leyte, and Manicani, Philippines.

92nd Bat. Formed 5-43 at Camp Peary, VA. Served at Oahu, Kauai, Saipan, and Tinian.

93rd Bat. Commissioned 11-10-43 at Camp Peary, VA. Served at Russell Is., Green Is., Solomons, Samar and Guiuan.

94th Bat. Formed 5-43. Served at Pearl Harbor, Apra Harbor, Guam and Marianas Islands.

95th Bat. Formed 11-43 at Camp Peary, VA. Served at Apamama, Gilbert Is., Roi-Namur and Iwo Jima.

96th Bat. Activated 6-12-43. Served at Terceira, Azores, Manicani Is., Samar, Philippines and Guiuan.

97th Bat. Formed 6-18-43 at Camp Peary, VA. Served at Londonderry, No. Ireland, London, Dunkeswell, Exeter, Heathfield, Lough Neagh, Plymouth, Salcombe, Dartmouth, Teignmouth, Southampton, Hants, Portland-Weymouth, Dorset, Fowey, Falmouth, Cornwall, Milford-Haven, Wales and Rosneath, Scotland.

98th Bat. Commissioned 6-30-43 at Camp Peary, VA. Served at Waiawa Gulch, Tarawa, Gilbert Is., Cora Is., Helen Is., and Maui.

99th Bat. Activated 6-24-43 at Camp Peary, VA. Served at Hawaii, Johnston Is., French Frigate Shoals, Canton Is., Angaur, Palau Is., Moanalua Ridge and Samar.

100th Bat. The "Century Battalion" was commissioned 7-1-43 at Camp Peary, VA. Served at Majuro Atoll, Marshall Is., Angaur, Palau Is., Guiuan and Samar.

101st Bat. Commissioned 8-13-43 at Camp Endicott, RI. Served at Saipan and Okinawa.

102nd Bat. Commissioned 8-19-43 at Camp Endicott, RI. Served at Finschaven, New Guinea, Hollandia, Subic Bay and Luzon, Philippines.

103rd Bat. Formed 10-15-43 at Camp Peary, VA. Served at Ojai, Mira Loma, Oxnard, CA, San Clemente, St. Nicholas Is. and Guam.

104th Bat. Formed at Camp Peary, VA. Arrived at Milne Bay on 2-2-44. Also served at Gamadodo, Los Negros, Australia, Saul Port and Leyte, Philippines.

105th Bat. Formed 8-43 at Camp Peary, VA. Served at San Clemente Is., Milne Bay, New Guinea, Hilimoi, Gamadodo, Stringer Bay, Tacloban, Leyte, San Pedro Bay, Anabong, San Antonio and Samar, Philippines, Talosa, Guiuan, Balingaga, and Osmena.

106th Bat. Formed 10-19-43 at Camp Peary, VA. Served at Pearl Harbor, Iwo Jima, Port Hueneme and Ie Shima.

107th Bat. Formed 7-43 at Camp Peary, VA. Served at Iroquois Pt., Oahu, Kwajalein Atoll, Ebeye Is., Bigej, Marshall Is. and Tinian.

108th Bat. Formed 8-43 at Camp Peary, VA. Served at Rosneath, Scotland, Plymouth, Netley, Normandy, and Tilbury, England.

109th Bat. Formed 7-43 at Camp Peary, VA. Served at Oahu, Kwajalein, Roi-Namur and Guam.

110th Bat. Commissioned 8-12-43 at Camp Peary, VA. Served at Port Hueneme, Eniwetok and Tinian.

111th Bat. Commissioned 9-43 at Camp Peary, VA. Served at Plymouth, Falmouth, Dartmouth, Swansea, Normandy, Calicoan, Samar, Mindanao, Tarakan, Brunei Bay, and Balikpapan, Borneo.

112th Bat. Formed in 1943 at Camp Peary, VA. Served at Tinian and Okinawa.

113th Bat. Formed in 1943 at Camp Peary, VA. Served at Hollandia, Wrong Is., Amsterdam Is., Soemesoeme Is., Samar, Leyte Gulf, Mindoro Is.

114th Bat. Formed at Camp Peary, VA, Summer, 1943. Served at Rosneath, Scotland, Cherbourg, France, Nantes, Pontivy and Attu, AK.

115th Bat. Formed 1943 at Camp Peary, VA. Served at Milne Bay, New Guinea, Brisbane, Australia, Luzon and Subic Bay, Philippines.

116th Bat. Arrived at Pearl Harbor on 3-5-44. Served at Oahu, Camp Tarawa on the island of Hawaii, and Japan.

117th Bat. Activated at Camp Peary, VA in 1943. Served in Oahu and Saipan.

118th Bat. Commissioned at Camp Peary, VA, Summer, 1943. Served at Gamadodo, Milne Bay, Mindanao, Zamboanga and Subic Bay, Philippines.

119th Bat. Formed at Camp Peary, VA, Summer, 1943. Served at Milne Bay, Hollandia, Aitape, Wopde Is., and Manila.

120th Bat. Formed at Casablanca, No. Africa on 2-19-43. Served at Oran, Arzero, Casablanca, Port Lyautey, Algiers, Jura, Span, Palermo, Sicily, and Termini.

121st Bat. Departed from San Diego, CA, on 1-8-44. Served at Roi-Namur Is., Maui, Saipan, and Tinian. The battalion was awarded the Presidential Citation for its construction and combat operations while attached to the 4th Marine Division.

122nd Bat. Formed at Camp Peary, VA, October, 1943. Served at Milne Bay, New Guinea, Gamadodo, Hollandia and Samar.

123rd Bat. Commissioned at Camp Peary, VA, Summer, 1943. Served at Moanalua Ridge, Pearl Harbor, Midway, Barber's Point, Oahu and Samar.

124th NCB Formed at Camp Parks, CA. Maintained the fleet base and headquarters for the 17th Naval District centered at Adak, Aleutian Islands.

125th Bat. Departed from Port Hueneme on January, 1944, for Hawaii. Served at Okinawa and Nakagusuku, Japan.

126th Bat. Formed at Camp Peary, VA, 1943. Served at Engebi, Eniwetok Atoll, Japtan, Parry and Hawthorne Is. in the Marshalls and Okinawa.

127th Bat. Departed from Port Hueneme on 5-1-44, Served in Hawaii, Leyte-Samar area of the Philippines and Japan.

128th Bat. Formed at Camp Peary, VA. Assigned to permanent pontoon operating base at Guam. Various detachments were sent on amphibious operations.

129th Bat. Activated at Camp Peary, VA. Arrived at Oahu on 4-1-44. Served in the Leyte-Samar area of the Philippines.

130th Bat. Formed at Camp Peary, VA, and shipped to Pearl Harbor during February, 1944. Served at Saipan and Okinawa in part as malaria control teams.

131st Bat. Formed at Camp Peary, VA, on 9-2-43. Received advanced training at Camp Endicott, RI, and served at Camp Parks, CA.

132nd Bat. Formed at Camp Peary, VA, on 10-12-43. Was inactivated at Camp Parks, CA, on 10-29-43.

133rd Bat. Formed at Camp Peary, VA, on 9-17-43. The full battalion landed at Iwo Jima on D-Day with the initial assault waves of the 4th Marine Division. The battalion suffered severe casualties and distinguished itself in both front line construction and combat. Navy Unit Commendation Award was received Sept., 1945. Recommissioned 8-12-66 at Gulfport, MS. The battalion received a second Unit Commendation for its 1966 deployment in Vietnam.

134th Bat. Activated in the field on 6-1-45 and formed from the personnel making up the motor pool (trucks) on Guam.

135th Bat. Activated at Camp Peary, VA. Departed for Pearl Harbor on 5-23-44. Served at Tinian and Okinawa.

136th Bat. Commissioned at Camp Peary, VA, during Sept., 1943. Served at Pearl Harbor, Guam and Yokosuka, Japan.

137th Bat. Formed at Camp Endicott, RI. Shipped to Okinawa 8-1-45 to provide peacetime services.

138th Bat. Formed at Attu on 2-1-44. Also served at Adak, Alaska. Inactivated June, 1945.

139th Bat. Commissioned at Camp Endicott, RI. Arrived at Port Hueneme on 2-6-45. Served at Okinawa.

140th Bat. Departed from Port Hueneme on 5-2-44. Served at Manus Is., Ponam Is., and Pityilu Is.

141st Bat. Commissioned at Camp Peary, VA, 10-43. Arrived at Pearl Harbor on 3-16-44 and served at Iroquois Pt., Oahu and Kwajalein.

142nd Bat. Departed from Port Hueneme on 6-4-44. Served at Leyte-Samar area of the Philippines.

143rd Bat. Commissioned 12-16-44 at Davisville, RI. Served at Samar, Philippines.

144th Bat. Arrived at Guam 3-18-45. Battalion located there on V-J Day.

145th Bat. Arrived at Okinawa on 5-1-45.

146th Bat. Relocated from Iceland to England during 2-44. Served at Plymouth until the Normandy Invasion. Operated at Omaha and Utah Beaches, Cherbourg and Okinawa.

147th Bat. Formed at Davisville, RI, 1945. Served at Okinawa.

148th Bat. On 5-20-45 the battalion departed for amphibious training at Morro Bay, CA. Served at Okinawa.

301st Bat. Formed as a Harbor Reclamation Battalion. First arrived at Pearl Harbor during 4-44. Served at Midway, Iroquois Pt, Oahu, Roi-Namur, Guam, Kwajalein, Saipan, Peleliu, Tinian, Iwo Jima and Okinawa.

302nd Bat. Formed at Pearl Harbor on 8-26-44. Served at Russell Is., Peleliu, Angaur, Leyte and Luzon, Okinawa, Oahu and Japan.

Special Battalion

1st Special Bat. Formed at Camp Peary, VA, Dec., 1942. Served at New Zealand, Guadalcanal, New Caledonia and New Hebrides.

2nd Special Bat. Formed at Camp Peary, VA, Jan., 1943. Served at Noumea, New Caledonia, Guadalcanal, and Guam.

3rd Special Bat. Formed Camp Peary, VA, on 1-24-43. Served at New Hebrides and Okinawa.

4th Special Bat. Formed at Camp Peary, VA, Feb., 1943. Served at New Caledonia, Guadalcanal, Tasafaronga and Okinawa.

5th Special Bat. Formed at Camp Peary, VA, Jan. 30, 1943. Served at Alaska, Emeryville, CA, Leyte and Samar, Philippines.

6th Special Bat. Arrived at Fiji Islands on 5-15-43. Served at Guadalcanal, Bougainville, Treasury Is., Ulithi, Oahu, Vella Lavella, and Russell Is.

7th Special Bat. Formed at Camp Peary, VA, 1943. Served at Alaska and Clatskanie, Oregon.

8th Special Bat. Formed at Davisville, RI, 1943. Served at Alaska, Clatskanie, OR and Port Hueneme, CA.

9th Special Bat. Formed at Camp Peary, VA, April, 1943. Served at Guadalcanal, Russell Is., Tulagi, Sasavele, Bougainville, Green Is., Russell Is. and Solomon Is.

10th Special Bat. Formed at Camp Endicott, RI. Departed for Pearl Harbor on 2-23-44. Served at Midway and Oahu.

11th Special Bat. Departed for Noumea, New Caledonia on 10-30-43. Served at Russell Is., Guadalcanal and Okinawa.

12th Special Bat. Arrived at Russell Is. on 1-7-44. Also served at Okinawa.

13th Special Bat. Arrived at Pearl Harbor on Nov. 27, 1943. Also served at Guam.

14th Special Bat. Arrived at Pearl Harbor, Oct., 1943. Served at Funafuti, Iroquois Pt., Tarawa, Marshall Is., Gilbert Is., Kwajalein, Eniwetok, Majuro and the Philippines.

15th Special Bat. Formed at Camp Peary, VA in 1943. Served at Gamadodo, New Guinea, Hollandia, Kwajalein and Marshall Islands.

16th Special Bat. Arrived at Pearl Harbor on 1-1-44. Served at Eniwetok and Guam.

17th Special Bat. Activated 9-19-43. Served at Russell Is., Emirau Is., Ulithi, Peleliu, Anguar, Guam and Leyte.

18th Special Bat. Arrived Oahu on May 1, 1944. Served at Pearl Harbor, Honolulu, Ulithi, Leyte and Tacloban.

19th Special Bat. Commissioned at Camp Peary, VA, 1943. Served at Finschaven, New Guinea, Biak Is. and Hollandia.

20th Special Bat. Formed at Camp Peary, VA, 1943. Served at Manus.

21st Special Bat. Formed at Camp Peary, VA, 1944. Served at Manus, Subic Bay and Luzon, Philippines.

22nd Special Bat. Departed for Manus Is, Admiralties on April 7, 1944.

23rd Special Bat. Served at Iwo Jima and Okinawa.

24th Special Bat. Served in New Guinea and Philippines.

25th Special Bat. Served at Milne Bay, New Guinea and Gamadodo.

26th Special Bat. Served at Pearl Harbor, Honolulu and Red Hill, Oahu.

27th Special Bat. Served at Tinian and Okinawa.

28th Special Bat. Served at Leyte and Samar area of the Philippines.

29th Special Bat. Formed at Camp Peary, VA, 1944. Served on the island of Guam.

30th Special Bat. Served at Rosneath, Scotland, Plymouth, England and Leyte-Samar area of the Philippines.

31st Special Bat. Served at Saipan.

32nd Special Bat. Served at Leyte-Samar area of the Philippines.

33rd Special Bat. Commissioned at Davisville, RI, August, 1944. Served at Milne Bay, New Guinea, Guiuan and Leyte-Samar.

34th Special Bat. Served from late 1944 to the end of the war at Guam.

35th Special Bat. Formed at Davisville, RI, 1944. Served at Pearl Harbor through the end of WWII.

36th Special Bat. Commissioned at Port Hueneme, CA, Jan. 20, 1945. Served at Okinawa.

37th Special Bat. Formed at Davisville, RI, 1945. Served at Pearl Harbor through the end of the war.

38th Special Bat. Served as stevedore replacement pool at Port Hueneme during Summer, 1945.

41st Special Bat. Served at Hollandia, New Guinea.

CONSTRUCTION BATTALION MAINTENANCE UNITS

CBMU 501 Served at New Zealand and Russell Is.

CBMU 502 Served at Wellington, New Zealand, Vella Lavella, Emirau, Manus and Guam.

CBMU 503 Served at Fiji, Russell Is. and Peleliu.

CBMU 504 Served at Wallis Is., Upolu Is, American Samoa, New Caledonia and Guam.

CBMU 505 Served at Upolu Is., Tulagi and Saipan.

CBMU 506 Served at Tutuila, Funafuti, Tongatabu, Samoa, New Caledonia and Guam.

CBMU 507 Served at St. Thomas, Virgin Is. and Puerto Rico.

CBMU 508 Served at Dutch Harbor, Alaska.

CBMU 509 Served at Amchitka, Adak, Tanga, Kiska, Ie Shima, Ryukyus, and Okinawa.

CBMU 510 Served at Cold Bay, Dutch Harbor, Otter Point and Adak. Second tour served at Saipan.

CBMU 511 Served at Efate, New Hebrides, Tongatabu, Wallis Is., and Guam.

CBMU 512 Served at Sitka, Alaska and Leyte-Samar area of the Philippines.

CBMU 513 Served at Oran, Algeria and Bizerte.

CBMU 514 Served at Iceland from Summer, 1944 through end of war.

CBMU 515 Arrived at Guadalcanal Nov., 1943. Also served at Kwajalein, Marianas, Saipan and Guam.

CBMU 516 Served at San Juan, Puerto Rico.

CBMU 517 Served at Funafuti, Russell Is., Guiuan, Samar and Ulithi.

CBMU 518 Served at Guadalcanal. Inactivated 8-10-45.

CBMU 519 Served at Noumea, New Caledonia until end of war.

CBMU 520 Served at Guadalcanal.

CBMU 521 Served at Tulagi and Okinawa.

CBMU 522 Served at Barber's Point, Oahu.

CBMU 523 Served at Iroquois Point, Oahu.

CBMU 524 Served at Midway, Eastern Is., and Sand Is.

CBMU 525 Served at Argentia, Newfoundland.

CBMU 526 Served at Argentia in combination with CBMU 525.

CBMU 527 Served at Palmyra Is.

CBMU 528 Served at Milne Bay, New Guinea.

CBMU 529 Served at Milne Bay.

CBMU 530 Served at Midway and Ewa, Oahu.

CBMU 531 Served at Midway.

CBMU 532 Served at Guadalcanal, Russell Is., Anguar and Guam.

CBMU 533 Served at Guadalcanal.

CBMU 534 Served at Espiritu Santo, New Hebrides, Noumea, New Caledonia and Okinawa.

CBMU 535 Served at Espiritu Santo.

CBMU 536 Served at Noumea, New Caledonia.

CBMU 537 Served at Noumea.

CBMU 538 Served at Espiritu Santo and Noumea.

CBMU 539 Served at Espiritu Santo, and Efate, New Hebrides.

CBMU 540 Served at Bermuda.

CBMU 541 Served at Espiritu Santo and Okinawa.

CBMU 542 Served at Espiritu Santo.

CBMU 543 Served at Finschaven, New Guinea and Subic Bay.

CBMU 544 Served at Brisbane, Australia and Leyte-Samar, Philippines.

CBMU 545 Served at Finschaven, New Guinea, Milne Bay and Hollandia.

CBMU 546 Served at Cairns, Australia, Port Moresby, Milne Bay, Hollandia and Palawan.

CBMU 547 Served at Attu, Alaska.

CBMU 548 Served at Gamadodo, Milne Bay, Hollandia and Manila.

CBMU 549 Served at Tarawa and Kwajalein.

CBMU 550 Served at Efate, New Hebrides, Noumea, New Caledonia and Banika, Russell Is.

CBMU 551 Served at Bermuda. Was combined with CBMU 540 Dec., 1943.

CBMU 552 Served at Nukufetau, Ellice Is., Green Is. and Hollandia, New Guinea.

CBMU 553 Served at Nanumea, Ellice Is., Green Is. and Leyte-Samar, Philippines.

CBMU 554 Served at Johnston Island.

CBMU 555 Served at Panama, Salinas, Balboa, Corinto, Taboga and Barranquilla, Canal Zone.

CBMU 556 Served at Attu, Alaska.

CBMU 557 Served at Apamama, Gilbert Is. and Guam.

CBMU 558 Served at Finschaven and Hollandia, New Guinea.

CBMU 559 Served at NOB, Trinidad (merged with CBMU 560), Also small detachments to Curacao and British Guiana.

CBMU 560 Served at Trinidad (see CBMU 559).

CBMU 561 Served at Munda and Ondonga, New Georgia Is. and Manus.

CBMU 562 Served at Hilo, Hawaii.

CBMU 563 Served at Naval Air Station, Kahului, Maui, Hawaiian Is.

CBMU 564 Served at Keehi Lagoon, Oahu, NAS, Barking Sands, Kauai and NAS, Honolulu.

CBMU 565 Served at Milne Bay, new Guinea, Morotai Is. and Tarakan, Borneo.

CBMU 566 Served at NOB, Casablanca, No. Africa, Agadir and Pt. Lyautey.

CBMU 567 Served at Palermo, Sicily, Salerno and Naples, Italy.

CBMU 568 Served at Munda, New Georgia and Samar, Philippines.

CBMU 569 Served at Treasury Is. and Samar, Philippines.

CBMU 570 Served at Oceanside, CA and Guam.

CBMU 571 Served Russell Is. and Peleliu.

CBMU 572 Served at Russell Is.

CBMU 573 Served at Russell Is. until end of war (with CBMU 572).

CBMU 574 Served at NAS, Pearl Harbor.

CBMU 575 Served at NAS, Puunene, Maui.

CBMU 576 Served at Attu, Alaska.

CBMU 577 Served at Tarawa, Eniwetok and Engebi.

CBMU 578 Formed at Arzew, Algeria, Nov. 1943. Second tour to Okinawa.

CBMU 579 Served at Nettuno and Anzio, Italy, Arzew, Algeria and Okinawa.

CBMU 580 Served at Segi, New Georgia Is., Munda, Russell Is. and Okinawa.

CBMU 581 Served at Pearl Harbor.

CBMU 582 Served at Torokina and Samar.

CBMU 583 Served at England.

CBMU 584 Served at Dunkeswell, England.

CBMU 585 Served at Milne Bay, Manus and Manila.

CBMU 586 Served at Torokina, Bougainville, Tacloban and Leyte.

CBMU 587 Served at Manus and Pityilu Is.

CBMU 588 Served at Canton Is.

CBMU 589 Served at Port Hueneme, CA.

CBMU 590 Served at Roi, Marshall Is.

CBMU 591 Served at Majuro, Marshall Is.

CBMU 592 Served at Eniwetok.

CBMU 593 Served at Tinian, Guam and Saipan.

CBMU 594 Served at Engebi, Marshall Is. and Guam.

CBMU 595 Served at Pearl Harbor and Saipan.

CBMU 596 Served at Kaneohe Bay, Oahu and French Frigate Shoals.

CBMU 597 Served at Tinian.

CBMU 598 Served at Leyte, Philippines.

CBMU 599 Served at Pearl Harbor.

CBMU 600 Served at Pearl Harbor (maintaining three hospitals and a Naval Medical Supply Depot).

CBMU 601 Served at Ebeye, Marshall Is.

CBMU 602 Served at Ulithi, Guam and Tokyo.

CBMU 603 Served at Ulithi.

CBMU 604 Served at Port Hueneme and Camp Parks, CA.

CBMU 605 Served at Borneo and Biak.

CBMU 606 Served at Milne Bay, Luzon, Lingaylu and Clark Field, Manila.

CBMU 607 Served at Berlin Is., Marshalls, Tarawa and Kwajalein.

CBMU 608 Served at Eniwetok, Marshalls.

CBMU 609 Served at Manus and Mindoro.

CBMU 610 Served at Manus.

CBMU 611 Served at Arzew, Algeria, Toulon and Marseilles, France.

CBMU 612 Served at Manus.

CBMU 613 Served at the Azores.

CBMU 614 Served at Saipan.

CBMU 615 Served at Okinawa.

CBMU 616 Served at Saipan.

CBMU 617 Served at Okinawa. Worked at Yonton and Chimu airfields until end of the war.

CBMU 618 Served at Camp Parks and Port Hueneme, CA, and Okinawa.

CBMU 619 Served at Guam.

CBMU 620 Served at Pearl Harbor and Iwo Jima.

CBMU 621 Served at the Admiralties and Manus.

CBMU 622 Served at Leyte, Guiuian and Samar.

CBMU 623 Served at Leyte-Samar, Philippines.

CBMU 624 Served at Pearl Harbor and Okinawa.

CBMU 625 Served at Acorn Tra Det, Port Hueneme and Okinawa.

CBMU 626 Served at NOB, Oran and Arzew, Algeria.

CBMU 627 Formed Cherbourg, France, Nov., 1944. Inactivated July 28, 1945.

CBMU 628 Formed at Le Havre, France, Nov., 1944.

CBMU 629 Formed at Le Havre, France. Served at Orly Field.

CMBU 630 Served at Okinawa.

CBMU 631 Served at Okinawa.

CBMU 632 Served at Okinawa through end of the war.

CBMU 633 Served at Okinawa.

CBMU 634 Served at Kodiak, Alaska, through end of the war.

CBMU 635 Served Dutch Harbor, Alaska.

CBMU 636 Served at Bremerhaven, Germany.

Construction Battalion Detachments

CBD 1001 Served at Guirock, Scotland and Freetown, Africa.

CBD 1002 Served at Scotland and Freetown, Africa.

CBD 1003 Served at Noumea, New Caledonia with First Marine Amphibious Corps.

CBD 1004 Served at Argentia, Newfoundland.

CBD 1005 Served at Arzew, Algeria, Bizerte, and Maddelena, Sardinia.

CBD 1006 Served at Bizerte, No. Africa, Sicily, Exeter, Plymouth, Falmouth, Dartmouth and Southampton, England. Placed pontoon causeways at Utah ad Omaha beaches, Normandy.

CBD 1007 Served at Espiritu Santo, New Hebrides.

CBD 1008 Served at Florida Is. (near Solomons) and Tulagi.

CBD 1009 Served at Russell Is.

CBD 1010 Served at Tulagi, Solomons and Guam.

CBD 1011 Served at Fort Pierce, Florida.

CBD 1012 Served at Panama, Balboa, Canal Zone, Ecuador, Nicaragua and Honduras.

CBD 1013 Served at Espiritu Santo, New Hebrides.

CBD 1014 Served at Espiritu Santo (inactivated 3-44).

CBD 1015 Served at Espiritu Santo (inactivated 4-44).

CBD 1016 Served at Guadalcanal (inactivated 4-44).

CBD 1017 Served at Oran, No. Africa, Kodiak and Cold Bay, Alaska.

CBD 1018 Served at Attu, Alaska.

CBD 1019 Served at Noumea, Espiritu Santo, and Tulagi.

CBD 1020 Served at Formed at Camp Peary, VA, 1943. Assigned to Spare Parts Control Units (inactivated late 1943).

CBD 1021 Formed as an Equipment Maintenance and Repair Detachment (Duty cancelled).

CBD 1022 Served at Adak, Alaska and Samar, Philippines.

CBD 1023 Served at Noumea, New Caledonia, Milne Bay and Gamadodo.

CBD 1024 Served at Milne Bay, Leyte-Samar and Guiuan.

CBD 1025 Served at Gulfport, Mississippi.

CBD 1026 Formed at Davisville, RI (9-43), Inactivated (10-43).

CBD 1027 Served at Solomon Islands and Little Creek, Virginia.

CBD 1028 Served at Quoddy Village, Maine.

CBD 1029 Served at Espiritu Santo.

CBD 1030 Served at Central Spare Parts Depot, Joliet, Illinois.

CBD 1031 Served at Solomon Islands (inactivated 2-7-45).

CBD 1032 Served at Solomon Islands.

CBD 1033 Served at Trinidad and Pearl Harbor.

CBD 1034 Served at Majuro and Japton, Marshall Is.

CBD 1035 Served at Marshall Is., Saipan and Tinian.

CBD 1036 Served at Tinian.

CBD 1037 Formed at Port Hueneme (12-43). Inactivated June 6, 1945.

CBD 1038 Served at Pearl Harbor.

CBD 1039 Served at Pearl Harbor and Marshall Is.

CBD 1040 Served at Bizerte, Naples-Salerno, Italy, Marseille and Toulon, France and Oran, No. Africa.

CBD 1041 Served at Pearl Harbor.

CBD 1042 Served at Pearl Harbor (Inactivated 5-45).

CBD 1043 Served at Pearl Harbor. (Inactivated personnel transferred to 302nd Bat., 8-44).

CBD 1044 Served at Kwajalein, Eniwetok, Guam, Ulithi, Nakagusuku, Okinawa.

CBD 1045 Served at Normandy on D-Day, Toulon and Marseilles, France, Calvi and Ajaccio, Corsica.

CBD 1046 Served at Espiritu Santo and Guam.

CBD 1047 Formed at Davisville, RI (2-44). Inactivated (12-44).

CBD 1048 Served at Plymouth, England.

CBD 1049 Served at Naval Amphibious Supply Base, Exeter, England.

CBD 1050 Served at Manus, Admiralties.

CBD 1051 Served at Manus.

CBD 1052 Served at Adak, Alaska.

CBD 1053 Served at Guam, Samar and Manicani Is.

CBD 1054 Served at Russell Is.

CBD 1055 Served at Espiritu Santo and Guam.

CBD 1056 Served at Guadalcanal, Espiritu Santo and Noumea.

CBD 1057 Formed at Gulfport, MS (10-44). Inactivated soon afterward.

CBD 1058 Served at Point Barrow, Alaska.

CBD 1059 Served at Guam.

CBD 1060 Served at Pearl Harbor, Eniwetok, Ulithi and Okinawa.

CBD 1061 Formed at Davisville, RI (7-16-44). Served aboard *USS Allegan.*

CBD 1062 Formed at Davisville, RI (7-16-44). Served aboard *USS Apponoose.*

CBD 1063 Served at Manus, Admiralties and Manila, Philippines.

CBD 1064 Served at Manus and Guam.

CBD 1065 Served Tinian and Marianas.

CBD 1066 Served at Manus and Leyte-Samar, Philippines.

CBD 1067 Served at Samar, Manus and Guiuan.

CBD 1068 Served at Kwajalein. Inactivated 8-45.

CBD 1069 Served at Saipan in part operating a Seabee dredge.

CBD 1070 Served at Guam.

CBD 1071 Arrived at Guam March 19, 1945. Inactivated August 8, 1945.

CBD 1072 Arrived at Guam November 23, 1944.

CBD 1073 Arrived at Guam March 19, 1945.

CBD 1074 N/A

CBD 1075 Formed at Port Hueneme on 10-27-44. Inactivated 9-5-45.

CBD 1076 Served at Dutch Harbor, Alaska.

CBD 1077 Served at Attu, Alaska.

CBD 1078 Served at Iwo Jima.

CBD 1079 Served at Okinawa.

CBD 1080 Served at Tinian, Marianas.

CBD 1081 Served at Okinawa.

CBD 1082 Served at Hollandia, New Guinea and Subic Bay, Philippines.

CBD 1083 Formed Camp Parks, CA, February, 1945. Soon inactivated.

CBD 1084 Served at Guam.

CBD 1085 Served at Hollandia, New Guinea.

CBD 1086 Served at Peleliu.

CBD 1087 Served at Okinawa.

CBD 1088 Served at Okinawa.

CBD 1089 Served at Tinian.

CBD 1090 Served at Guam.

CBD 1091 Served at Okinawa.

CBD 1092 Served at Milne Bay, New Guinea, Manus and Subic Bay, Philippines.

CBD 1093 Served at Saipan.

CBD 1094 Formed at Port Hueneme, July, 1945. Inactivated Sept. 15, 1945.

CBD 1101 Activated from personnel operating ABCD at Manus, Admiralty Is., September 12, 1945.

CBD 3050 Constructed the Seabee Camp at Quoddy Village, Maine. Inactivated January 19, 1945.

PONTOON ASSEMBLY DETACHMENTS

Pontoon Assembly Det. One Served at Noumea, Manus and Samar.

Pontoon Assembly Det. Two Served at Russell Is and Guam.

Pontoon Assembly Det. Three Served at Milne Bay and Leyte-Samar, Philippines.

Pontoon Assembly Det. Four Served at Hollandia, New Guinea and Leyte-Samar, Philippines.

Pontoon Assembly Det. Five Served at Guam.

NAVAL CONSTRUCTION BATTALIONS

NCB-104 Formed in 1950. Arrived at Camp McGill, Japan, at the outbreak of the Korean conflict. Later redesignated ACB-1.

AMPHIBIOUS CONSTRUCTION BATTALIONS

ACB-1 Organized 10-50 from NCB-104. Served at Inchon, Wolmi-Do, Red and Blue Beaches, and Yo Island in the Bay of Wonsan.

ACB-2 Activated 10-50. Served at Ascension Is., Dominican Republic, Greece, Haiti, Italy, Lebanon, Puerto Rico and Spain.

CONSTRUCTION BATTALION MAINTENANCE UNITS

CBMU-1 Activated July, 1952. Served at Korea and Japan. Inactivated October, 1953.

CBMU-101 Activated October, 1953. Served at Japan. Inactivated March, 1956.

MOBILE CONSTRUCTION BATTALIONS

MCB-1 Activated Davisville, RI, August, 1949. Served at Antarctica, Bermuda, Costa Rica, Cuba, Morocco, Newfoundland, Spain.

MCB-2 Activated Port Hueneme, CA, June, 1950. Served at Japan, Korea, Midway, Philippines and Samoa. Inactivated August, 1956.

MCB-3 Served at Alaska, Chi Chi Jima, Guam, Hawaii, Iwo Jima, Okinawa, Philippines, Thailand and Yap.

MCB-4 Served at Bahamas, Bermuda, Cuba, Ecuador, Haiti, Morocco, Newfoundland, Puerto Rico, Scotland, Trinidad and Spain.

MCB-5 Served at Alaska, Eniwetok, Guam, Hawaii, Midway, Okinawa and Philippines.

MCB-6 Served at Antarctica, Antiqua, Bermuda, Cuba, Greece, Morocco, Newfoundland, Puerto Rico and Spain.

MCB-7 Served at Barbados, Cuba, El Salvador, Ethiopia, Morocco, Newfoundland, Puerto Rico, Sicily, Scotland, Spain and Trinidad. Inactivated August, 1970.

MCB-8 Served at Antarctica, Bermuda, Cuba, Greece, Morocco, Newfoundland, New Zealand, Spain, and Turkey.

MCB-9 Served at Alaska, Guam, Hawaii, Kwajalein, Marcus Is., Midway, Okinawa, Philippines, Spain and Taiwan.

MCB-10 Served at Alaska, Anguar, Antarctica, Canton Is., Guam, Kwajalein, Midway, Okinawa, Philippines, Saipan and Ulithi.

MCB-11 Served at Alaska, Guam, Kwajalein, Midway, Okinawa and Philippines.

VIETNAM WAR-ERA BATTALIONS

MCB-1 Formed August, 1949, at Davisville, RI. By November, 1965, served at Danang. Later served at Phu Bai, Vietnam, Roosevelt Roads, Puerto Rico, Diego Garcia, Chagos Is., Guam, Marianas Is. and Rota, Spain.

MCB-3 Formed July, 1950. By 1965 served at Danang, Chu Lai, Phu Bai-Gia Le, Okinawa, Ryukyu Is., Marianas Is. and Puerto Rico.

MCB-4 Formed 1950. Joined Pacific Fleet Battalions November, 1965. Served at Chu Lai, Danang, Okinawa, Ryukyu Is., Puerto Rico, Diego Garcia, Chagos Is. and Guam.

MCB-5 Activated 1951. By April, 1965, served at Danang, Dong Ha, Bien Hoa, Guam, Marianas Is., Okinawa, Ryukyu Is., Nam Phong, Thailand, Okinawa, Ryukyu Is. and Roosevelt Roads, Puerto Rico.

MCB-6 Served at Marathon, Greece, Rota, Spain, Antarctica, Guantanamo Bay, Cuba, Danang, Chu Lai and Roosevelt Roads, Puerto Rico. Inactivated November 17, 1969.

MCB-7 Served at Phu Bai, Danang, Dong Ha and Chu Lai.

MCB-8 Served at Danang, Chu Lai and Phu Bai-Gia Le.

MCB-9 Served at Danang, Okinawa and Ryukyu Is.

MCB-10 Formed October, 1952. By 1965 served at Chu Lai, Danang, Okinawa,, Ryukyu Is., Phu Bai-Gia Le, Quang Tri, Camp Evans, Vietnam, Okinawa, Ryukyu Is., Rota, Spain, Roosevelt Roads, Puerto Rico and Diego Garcia, Chagos Is.

MCB-11 Formed July, 1953. Served at Guam, Marianas Is., Okinawa, Ryukyu Is., Danang, Dong Ha, Quang Tri and Camp Evans, Vietnam.

MCB-12 Reserve battalion activated at Gulfport, MS, May, 1968. Served at Danang.

MCB-22 Activated May, 1968. Served at Danang.

MCB-40 Re-activated February, 1966. Served at Chu Lai, Hue, Phu Bai, Roosevelt Roads, Puerto Rico, Diego Garcia, Guam, Rota, Spain, Okinawa and Ryukyu Is.

MCB-53 Re-activated June, 1967. Served at Danang, Vietnam.

MCB-58 Re-activated March, 1966. Served at Danang and Chu Lai, Vietnam.

MCB-62 Served at Phu Bai, Danang, Dong Ha, Vietnam, Roosevelt Roads, Puerto Rico, Diego Garcia, Guam and Rota, Spain.

MCB-71 Served at Chu Lai, Vietnam, Roosevelt Roads, Puerto Rico, Guantanamo Bay, Cuba, Antarctica and Bermuda.

MCB-74 Served at Danang, Chu Lai, Bien Hoa, Vietnam, Roosevelt Roads, Puerto Rico, Diego Garcia and Guam.

MCB-121 Served at Hue, Phu Bai, Phu Bai-Gia Le and Danang, Vietnam.

MCB-128 Served at Danang and Quang Tri, Vietnam.

MCB-133 Served at Danang, Hue-Phu Bai, Phu Bai-Gia Le, Vietnam, Guam, Okinawa, Ryukyu Is. and Rota, Spain.

CONSTRUCTION BATTALION MAINTENANCE UNITS

CBMU 301. Formed at Port Hueneme, CA, March 31, 1967. Served at Dong Ha and Chu Lai, Vietnam. Disestablished October 30, 1970.

CBMU 302 Served at Cam Ranh Bay, Bien Hoa, Vietnam and Subic Bay, Philippines.

SEABEE UNITS IN 1987

NAVAL CONSTRUCTION REGIMENTS

20th NCR—Gulfport, MS.
31st NCR—Port Hueneme, CA.

AMPHIBIOUS CONSTRUCTION BATTALIONS

PH1BCB—Coronado, CA.
PH1BCB—Little Creek, VA.

NAVAL MOBLE CONSTRUCTION BATTALIONS

NMCB-1 Gulfport, MS.
NMCB-3 Port Hueneme, CA.
NMCB-4 Port Hueneme, CA.
NMCB-5 Port Hueneme, CA.
NMCB-7 Gulfport, MS.
NMCB-40 Port Hueneme, CA.
NMCB-62 Gulfport, MS.
NMCB-74 Gulfport, MS.
NMCB-133 Gulfport, MS.

CONSTRUCTION BATTALION MAINTENANCE UNIT

CBMU 302 Subic Bay, Philippines.

UNDERWATER CONSTRUCTION TEAMS

UCT-1 Little Creek, VA.
UCT-2 Port Hueneme, CA.

NAVAL CONSTRUCTION BATTALION UNITS

CBU-401 Naval Training Center, Great Lakes, IL.
CBU-402 Naval Air Station, Pensacola, FL.
CBU-403 U.S. Naval Academy, Annapolis, MD.
CBU-404 Naval Air Station, Memphis, TN.
CBU-405 Naval Air Station, Miramar, CA.
CBU-406 Naval Air Station, Lemoore, CA.
CBU-407 Naval Air Station, Corpus Christi, TX.
CBU-408 Naval Education Training Center, Newport, RI.
CBU-409 Naval Station, Long Beach, CA.
CBU-410 Naval Air Station, Jacksonville, FL.
CBU-411 Naval Station, Norfolk, VA.
CBU-412 Naval Station, Charleston, SC.
CBU-413 Naval Station, Pearl Harbor, HI.
CBU-414 Naval Submarine Base, Groton, CT.
CBU-415 Naval Air Station Oceana, Virginia Beach, VA.
CBU-416 Naval Air Station, Alameda, CA.
CBU-417 Naval Air Station, Whidbey Is., WA.
CBU-418 Naval Submarine Base, Bangor, Bremerton, WA.
CBU-419 Naval Training Center, Orlando, FL.
CBU-420 Naval Station, Mayport, FL.
CBU-421 Naval Station, Mare Is., CA.
CBU-422 Navy Yard, Wash., D.C.*

*Courtesy of Jeffrey R. Millet, Publishing Consultant/Editor, Taylor Publishing Co.

APPENDIX A

The SEABEE insignia—a flying bee . . . fighting mad! On his head he wears a sailor hat. In his forehand he carries a Tommy gun, in his midship hand, wrench, and in his aft hand, a hammer.
(1942 poster courtesy of A. Clark Fay, CM 1/C)

SAMPLE OF THE MANY SEABEE PATCHES

THE FIGHTING SEABEES

THEY were called Seabees — and had a number of other names to denote what an irregular bunch they were.

The Seabee nickname came from CB, the abbreviation for Construction Battalion, a civilian branch of the U.S. Navy hastily authorized in late December 1941. It was organized with plumbers, carpenters, truck drivers and construction workers who were soon doing their jobs under combat fire: "We build, we fight" was one of their mottoes.

Based on the unorthodox manner of their work, they earned other names—some printable—such as Confused Bastards. Their insignia was a flying bee wearing a sailor cap and carrying a Tommy gun, a wrench and a hammer.

Seabees were older than the average sailors—31—a fact that led to many jokes: "Don't pick on a Marine; he may be the son of a Seabee," sailors would say. But the Seabees' wartime accomplishments gave the jokes a foundation of respect and admiration. Wherever overseas bases, airfields and waterfront facilities were needed, the Seabees provided them in their typical "Can Do" spirit that became another of their mottoes.

WIDE WORLD

BUILDING & FIGHTING—***Wherever airfields or bases were needed, the Seabees were there.***

Their "Mulberries"—artificial harbors—were essential to the D-Day channel crossings. Their "magic box" pontoons made the landings at Salerno and Anzio possible. They unloaded 10,000 vehicles from Landing Ship Tanks (LSTs) in Sicily in 23 days. They leveled the mountains of Ascension Island and built a mile-long airstrip where most engineers said it couldn't be done.

On Okinawa, they built seaports with dock and cargo handling facilities, airfields and a seaplane base. On Guam, working under fire and using bulldozers as shields, they built and paved 100 miles of road in 90 days.

Adm. "Bull" Halsey called bulldozers one of the most decisive weapons of the Pacific war, along with subs, radar and planes.

William Bradford Huie wrote about them in the book, *Can Do*, and their contributions to the war were made legendary in a John Wayne movie, *The Fighting Seabees.* □

APPENDIX B

The Song Of The SEABEES

Lyric by
SAM M. LEWIS

Music by
PETER DE ROSE

Dedicated to the SEABEES
Construction and Fighting Men
of the UNITED STATES NAVY

Printed for complimentary distribution by Bureau of Yards and Docks, United States Navy, by

ROBBINS MUSIC CORPORATION
799 SEVENTH AVENUE • NEW YORK

NAVY DEPARTMENT

BUREAU OF YARDS AND DOCKS

WASHINGTON, D. C.

JOIN THE NAVY

The Construction Regiment's "SEABEES"

The "SEABEES" are the men who enlist in the Construction Regiment of the U. S. Navy to build the advance and mobile bases outside the continental limits of the United States.

The "SEABEES" will be thoroughly trained in military tactics and when assigned to duty will be able to engage in combat should the occasion arise.

The U. S. Navy has opened enlistments, with ratings, for men with construction experience for enrollment in Class V-6 of the Naval Reserve for assignment to headquarters and construction companies in a Construction Regiment. These headquarters and construction companies are comprised of mechanics, carpenters, electricians, power plant operators, blacksmiths, metalsmiths, drillers, divers, wharfbuilders, etc. Acting appointments are made to persons between the ages of 17 and 50 in various ratings up to and including Chief Petty Officer, depending upon the age, experience and other qualifications of the personnel enlisted.

The enlistment period is for the duration of the war. The salaries for these enlistments range from $54.00 to $126.00 a month and include housing, food, clothing, transportation, medical and dental care, and other incidentals to which enlisted personnel are entitled.

APPLY TO YOUR NEAREST NAVY RECRUITING STATION FOR INFORMATION, or communicate with Bureau of Yards & Docks, Room 1305 Navy Building, Washington, D. C., for an Application for Enlistment form.

Here is a real OPPORTUNITY for two-fisted, red-blooded Americans to serve shoulder-to-shoulder with the combatant forces in the "SEABEES," the newest arm of Uncle Sam's Navy.

This recruitment poster was printed on the back of *THE SONG OF THE SEABEES* sheet music. The music itself was copyrighted in 1942 by the United States Navy, Bureau of Yards and Docks, Washington, D.C. Lyrics: We're the Seabees of the Navy, We can build and we can fight. We'll pave a way to victory and guard it day and night. And we promise that we'll remember The "Seventh of December." We're the Seabees of the Navy. Bees of the Seven Seas. The Navy wanted men. That's where we came in—Mister Brown and Mister Jones, the Owens, the Cohen's and Flynn. The Navy wanted more of Uncle Sammy's kin, so we all joined up and brother we're in to win.• Courtesy of USN, Bureau of Yards and Docks.

APPENDIX C
PEARL HARBOR'S LAST CASUALTY

Ken Ringle, Washington Post Staff Writer, December 2, 1994 (Reprinted by permission).

SCHOLARS URGE POSTHUMOUS CLEARING OF ADMIRAL'S NAME

Three scholar-historians of the Japanese attack on Pearl Harbor urged yesterday that the United States mark the 50th anniversary of the end of World War II next year [written 12-94] by rehabilitating the Pacific Fleet commander blamed for the December 7 disaster.

Citing what they described as newly declassified evidence unearthed in the National Archives more than half a century after the attack, the three said the stain on the name of Admiral Husband E. Kimmel deserves to be removed in the interest of both history and the nation.

"I have always believed history is a constant dialogue between the present and the past," said John Costello, author of a new book on the first American battles of World War II. "The case against Admiral Kimmel deserves to be reopened so that dialogue can continue."

Many historians have long been troubled by what happened to Kimmel, who was found guilty of dereliction of duty in connection with the attack, particularly for his failure to maintain long-range aerial reconnaissance around the Hawaiian islands in the face of increased warnings of Japanese aggression.

Kimmel partisans have maintained for decades that the admiral was railroaded after Pearl Harbor—made a scapegoat for intra-service squabbling, bureaucratic bungling and political decisions that denied him both the cryptographic intelligence necessary to evaluate Japanese intentions and the aircraft to thwart them.

Kimmel died in 1968, however, and previous efforts by his family to have his case reopened have been turned down by the Navy.

NAVAL HISTORICAL CENTER

Historians say Admiral Husband E. Kimmel was made the scapegoat for U.S. bungling at Pearl Harbor.

Most of the evidence cited at the Archives yesterday was incremental rather than dramatic—the sort that extends an existing paper trail rather than steering it off in a sharp new direction. But University of Florida historian Michael Gannon said there is already enough to make a difference.

He unveiled a letter from former Chief of Naval Operations C.A.H. Trost, reversing Trost's 1988 refusal to reopen the Kimmel case. After examining the new arguments, Trost writes, "I believe such action is owed to the admiral, to his sons and to the Navy. No mistake should be allowed to stand in this sensitive matter, and I personally disavow my unwitting support of one" in the past.

Seated in the audience at the National Archives yesterday were Edward Kimmel, 73, of Wilmington, Del., and Thomas K. Kimmel, 80, of Annapolis, who have worked tirelessly for years to clear their father's name.

"We think this is a good time to try again, with the war back in the news once more," Edward Kimmel said later. "I think the new Congress might be favorably disposed."

Costello and Gannon were joined on the Archives Theater stage by World War II submariner-author Edward L. Beach, who announced his own planned publication next spring of "an emotional rather than a factual" re-arguing of the Kimmel case. The three agreed that there were blunders enough in the first American hours of World War II without blaming Kimmel for the losses at Pearl Harbor.

Costello argued that Pearl Harbor was vulnerable primarily because political commitments to England and the Philippines had left Hawaii with too few planes to mount a proper watch. And he pointed out, as have scholars and historians ever since the war, that General Douglas MacArthur, then in the Philippines, was caught just as unprepared by the Japanese despite hearing of the Pearl Harbor attack, which had occurred eight hours earlier, yet never was reprimanded, as Kimmel was. MacArthur's entire air force was destroyed on the ground in, Costello says, "a far greater strategic disaster than Pearl Harbor . . . yet the general was allowed to redeem himself" though continued command and victories later in the war. Kimmel was cashiered.

In 1944, Gannon pointed out, an investigative board of naval officers exonerated Kimmel of the charges against him and recommended that his name be cleared. But its findings were overturned by Admiral Ernest J. King, then chief of naval operations, who later declared he never looked at the court record he reversed. Subsequent naval officers, civilian and military, have been reluctant to second-guess King's decision.

"We hope to be able to start a movement in this country," to assess both Kimmel's case and the similar case of Lieutenant General Walter C. Short, the U.S. Army's Hawaiian commander during the Japanese attack, Beach told the Archives audience. "It's not just a case of being fair: It's a case of being truthful to history."

Note: This author's relationship to Admiral Kimmel is more coincidental than real. In sharing the same last name my interest has been acute for years. However, for that same reason, I've had to go to extremes to remain objective, or to convince others to seek an objective reappraisal of this major event in American military history. My intent has never been to bash Franklin Delano Roosevelt as the responsible commander in chief, nor to believe that FDR was some bumbling incompetent without a clue as to what was happening militarily in the Pacific Ocean, or with the belligerent Empire of Japan.

Was it chance alone that FDR (a former Assistant Secretary of the Navy) escaped any carrier damage at the attack on Pearl Harbor? Was it chance that FDR failed to share any of the Enigma cryptographic reports with his Hawaiian commanders? Was it chance only that grossly inadequate fuel supplies were provided to maintain a constant 360 degree aerial perimeter of the islands and at least 1,000 miles out? Was it chance that the National Park Service didn't want newly developed radar towers atop Hawaiian mountain tops?

Why maintain an obsessive D.C. concern about sabotage at Pacific bases and yet have absolutely no intelligence of Imperial Japanese fleet movements? The motive, in this author's researched opinion, was that FDR and select others repeatedly sought an "incident" that would turn a rigorously anti-war nation around to preserve *western European culture and millions of lives (especially the British Isles)* from the death-grip of militant fascism. In warfare, supreme commanders have perhaps always had to weigh the sacrifice of a few to preserve the lives of others. It could be argued that at Pearl Harbor the loss of life was *more profound than anticipated* because advisors had grossly underestimated and racially stereotyped the Japanese because of their apparent bumbling in the quagmire of Manchuria, China, their "lower average stature," and a propensity to make "junk toys" out of old American beer cans that had been sold to them as scrap.

APPENDIX D

HISTORIANS SHIFT BLAME FOR PEARL HARBOR

Article by Calvin Woodward, Associated Press, Washington, D.C. Published in *The Sunday Oregonian,* December 4, 1994 (Reprinted by permission).

New studies assert that two U.S. military leaders weren't derelict in defending Hawaii

WASHINGTON [DC]—For 50 years two U.S. military leaders have been regarded as sleeping sentries in the Japanese attack on Pearl Harbor, men who ignored signs that the heart of the Pacific Fleet was in mortal danger.

But some historians and old military hands insist that Rear Admiral Husband E. Kimmel and Lt. General Walter C. Short were wrongly disgraced.

They are pressing efforts to vindicate the men, both cited during the war for dereliction of duty and both now dead.

Kimmel and Short were hardly sleeping, British author John Costello said last week at a National Archives lecture, joined by two like-minded historians.

Instead, he said, they were like "two guards who had been blind-folded by the lack of intelligence, told to face the wrong direction and inadequately armed to fend off an enemy attack."

U.S. officials refused as recently as last year to revisit the matter, despite the urgings of the Kimmel family, 32 retired admirals and two former chairmen of the Joint Chiefs of Staff.

Restoring their reputations, said historian Edward Beach, the author of the coming book, "Scapegoats," would "be a remedial act of justice in the name of the country."

At least one former senior military official has changed his mind about Kimmel. Admiral C.A.H. Trost, who recommended letting the issue rest when he was chief of naval operations in 1988, now says the case should be reopened.

"No mistake should be allowed to stand in this sensitive matter, and I personally disavow my unwitting support of one," Trost, now retired, wrote in October to Navy Secretary John Dalton.

Democratic Senator Joseph Biden of Delaware, where Kimmel's sons live, wrote to Defense Secretary William Perry in October, urging him to promote Kimmel posthumously to four-star admiral, the rank he normally would have achieved, had he not been forced to retire.

Perry hasn't replied, said a spokesman for Biden's office.

Kimmel, who was commander of the Pacific Fleet [actually commander of the U.S. Fleet] and Short, commander of the Army's Hawaii department, were blamed in official inquiries for playing down intelligence reports warning of Japan's intentions and for mounting inadequate reconnaissance.

The surprise attack Dec 7, 1941, killed 2,403 people, sank or damaged 18 warships, destroyed 188 planes and drew the United States into World War II.

Costello says his new book, "Days of Infamy," builds on previously reported evidence that the Hawaiian commanders were not given the full benefit of American intelligence.

The commanders indeed shrugged off some 11th-hour signals of trouble, he said, including reports that the Japanese Consulate in Honolulu had destroyed its codes—an act often foretelling war.

But he said they did so because a steady drumbeat of intelligence from Washington had pointed to the Philippines as the likely target of Japan's move against U.S. forces.

In the Philippines, Costello said, Gen. Douglas MacArthur had received "the full diplomatic picture of Japan's march toward belligerence" but did nothing.

Nine hours after attacking Pearl Harbor, Japan bombed Clark Field in the Philippines, destroying 18 bombers, 55 fighters and 33 other planes.

U.S. code-breakers had been reading Japanese diplomatic cables for months. But some historians say messages that could have been key to a successful Pearl Harbor defense were misread or ignored in Washington or kept from Hawaii.•

PEARL HARBOR—DECEMBER 7, 1941

One of the most remarkable combat photographs of the war was made at the exact moment the U. S. destroyer *Shaw* blew up during the Japanese attack on Pearl Harbor, Hawaii, December 7, 1941. Photographer unknown, probably an enlisted combat photographer.
U. S. Navy Photo

APPENDIX E

World War II Merchant Marines Will Be Honored, 50 Years Late

by Alice Cantwell, *The Journal of Commerce,* April 18, 1995, p. 1A-2A. Originally titled, "Another WWII Battle to End As Mariners Receive Their Medals." (Reprinted by permission).

Members of the U.S. merchant fleet were pressed into wartime service but denied the benefits of military veterans.

NEW YORK—Fifty years late, some World War II sailors finally will get their medals.

Some of the men are still alive to receive them: in other cases, widows or other survivors will accept medals issued to merchant marines who were pressed into service. Some of the sailors served in more than one theater of war, and some were prisoners of war.

Even now there is controversy over whether the government, which at last has agreed to issue the medals, will pay for them. There are hopes that the maritime industry will step in where the government won't.

Some 300 medals will be awarded on May 6, and the credit goes to Ray Pettersen, 39, a Vietnam War veteran who said he was embarrassed when he learned that merchant marines were denied wartime medals.

"Until two years ago I did nothing but scream about the Vietnam vet, and what he was due," said Pettersen, a post-traumatic-stress counselor who has spent much of his time working on behalf of homeless veterans. "But I found out there was a whole segment of vets who had been ignored. I felt guilty when I realized how quiet these guys had been."

Pettersen, a liaison officer for veterans' affairs at the Veterans Affairs Medical Center in Northport, N.Y., decided to use his knowledge of "how the system works" to help the merchant marines.

Even so, "there's been a lot of confusion" at the U.S. Department of Transportation over whether the government will cover the cost of minting the medals, which amounts to about $6 apiece, Pettersen said. Some men will receive more than one medal.

"They're saying they won't pay for them," he said, but he hopes that policy will change. A Transportation spokesman could not be reached for comment.

But just in case the policy isn't altered, Pettersen has found sponsors to cover the cost of the medals. He is still hearing from merchant marines who qualify for the medals and says he will need more sponsors.

The role of the merchant marines in World War II has been fraught with controversy.

Thousands of civilian seafarers crewed the lightly armed U.S.-flag cargo ships that carried supplies to U.S. troops and the Allies during World War II, and some 7,000 were lost at sea in enemy action, said Martin Skrocki, public information officer at the U.S. Merchant Marine Academy at Kings Point, N.Y.

Merchant marines shipped out of ports throughout the country to bring supplies to every theater of the war. Although they were classified as civilians, they were engaged in battle, and they had a higher casualty rate than any of the uniformed services except the Marines Corps.

Battle for benefits

After they returned home, the merchant marines had to battle more than 40 years for veterans' benefits because they were not considered members of the armed forces and so were not entitled to many of the benefits, such as those provided by the GI Bill, that covered other veterans.

Some GIs say that the merchant marines, who were paid to crew the ships, do not deserve equal recognition, Pettersen said.

To those vets, Pettersen asks: "How much is it worth to risk your life for your country or to die for your country?" The vets usually end up seeing his point, he said.

The government "waited too long to do what's right," Pettersen said, adding that the maritime industry, which also benefited from the men's work, should help them now.

"If the government isn't going to recognize these guys, let's do it for them," he said.

When the merchant marines finally won official status in 1988, it was too late to use the GI Bill to go to college or mortgage a first home, as many other veterans were able to do. After a hard-fought battle against red tape, the sailors are now entitled to full benefits, including health and burial benefits.

But some are still waiting for their medals.

"I don't know how they held out that long," said Pettersen, citing the psychological need to get the official recognition for wartime service and heroism. The accounts of what the young men went through impressed him deeply.

The merchant marines served aboard vessels that often were torpedoed by German warships disguised as cargo ships. Survivors spent days adrift in lifeboats in perilous seas.

"Some of these guys were 14 years old and signed on as cabin boys. Some were 15 and were able-bodied seamen. Then when they became old enough they joined the Army, Navy or Marines," he said.

A tribute to WWII merchant marine veterans will highlight a community day open house at the Kings Point Academy on May 6, 1995. Certificates will be presented if the costs of the medals are not covered by then.•

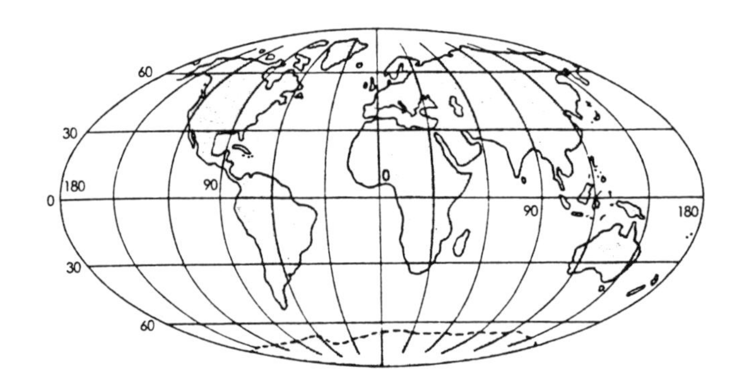

APPENDIX F

THE NAVY SEABEES: Their Early Days

by Nathan S. Raitt, Life Member, Seabee Veterans of America

Prior to the attack on Pearl Harbor on the morning of Sunday, December 7, 1941, the Navy, through its Civil Engineering Corps, was utilizing civilian labor under private contractors for the building of its advanced base facilities. By the 1st of December, 1941, there were about 70,000 civilians working on Navy projects outside of the continental United States. My impression was that this fact is one that was not generally known. At the time, I was employed by a leading department store in downtown Washington, D.C., where all manner of surmise was available (rumor mixed with facts) but even as a Naval Reservist, I had not heard a word.

The United States, under the leadership of President Franklin D. Roosevelt, was preparing facilities, presumably for our allies, but in retrospect, undoubtedly on the official assumption that sooner or later, the United States would be involved *(How little the small guy in the street knows what the powers that be have in mind).*

There were projects in Argentina, Iceland, North Ireland, Puerto Rico, St. Thomas, Trinidad, Jamaica, the Canal Zone and other parts of the Caribbean and the Atlantic. In the Pacific, projects were underway in the Galapagos, the Aleutians, Hawaii, Midway, Wake, Guam, Samoa, Palmyra, Johnson Islands, and the Philippines.

With the onset of war after December 7th, all of the workers in the Pacific were endangered, as proven by the taking of Wake Island by the Japanese in late December. At Wake, all of the surviving Marines and civilian workers of the original 350 Marines and 1,150 civilians were treated, or, perhaps more accurately, mistreated, as prisoners of war.

The Japanese attack had caught us totally unprepared with bases from which to fight back, and worse, we had no organizational set-up to construct such facilities as we would need. At Pearl Harbor there were 2,403 killed and 1,178 wounded.

Thus, the military concept of Construction Battalions was born. One source says that Admiral Ben Moreell, CEC, USN, secured the authority to enlist five companies of ninety-nine men each for construction duties in Iceland.

There seems to be some conflict in connection with dates and various actions. It is possible that at that hectic time most who were concerned were busy doing the essential things with little time for keeping detailed logs. It is noted that over the years since 1942, that people with different points of view have reported various dates as being the official one for the authorization of the 'Construction Battalions.'

Authority for the statement regarding Admiral Moreell, cited earlier, comes from *CAN DO,* the book by Lt. (jg) CEC, USNR William Bradford Huie, E. P. Dutton & Co., New York, © 1944. The introduction was written by Admiral Moreell. Huie's book states that the morale of the civilian workers in Iceland had deteriorated to the point that bonuses and higher wages were insufficient. The constant bad weather was depressing and the Icelandic people were unhappy with the occupation and were not very hospitable.

Huie almost contradicts himself in the same chapter by stating that, 'On December 28 (1941), when the first Naval Construction Regiment of 3,000 men was authorized.' It is possible that Admiral Moreell had been given unofficial permission relative to enlisting the first 395 men, and that *official authorization* for the regiment was perhaps given later.

Huie also states that enlistment must have begun immediately and that men began reporting in at the regular recruit Naval Training Station at Newport, RI, but that they never reached Iceland. Frantic demands from the Pacific required that Icelandic plans be put aside temporarily. There was an imperative call for a group of 250 men to leave for the South Pacific at once. The Navy met the demand by taking the men who had enlisted for duty in Iceland, and added some general service recruits. This group was organized at the Naval Air Station, Quonset Point, RI, on January 20, 1942. They were rushed to a waiting convoy at Charleston, SC, which sailed on January 27, 1942, for their destination on the island of Bora Bora in the South Pacific. I helped with the detail for this group—termed the 'Bobcats' from the operation code name. Sometime later they were merged with the 1st Construction Battalion; of which they were considered a detachment.

My memory may be a bit hazy after forty years [circa, 1983] since I recall that there were only 230 men in this group. It would be nice if, through one of our NSVA 'Hey, Buddy' ads, we could connect with one of this original 'Bobcat' group and verify the dates and number of men.

In a mimeographed booklet of the Navy Seabee Veterans of America, author unknown, prepared for an early convention in 1955, it is stated, 'The Seabees were made a permanent part of the Navy in 1946. However, the original commissioning of the 1st Seabee Battalion was on December 28, 1941, with ceremonies at Great Lakes and under the guiding hand of John Richard Perry who later became a rear admiral.'

In the 'ASK ANDY' column of the *Sarasota Herald Tribune*, Sept. 29, 1982, it is stated, 'The first *Battalion* of Seabees was authorized on January 5, 1942.' In 'Welcome Aboard,' a mimeographed publication for the 40th Anniversary held March 26-27, 1982, for the U.S. Naval Construction Battalion One by H. N. Wallin, Rear Admiral, CEC, USN, he states, 'By BuNav Circular Letter No. 1-42, dated January 5, 1942, the bureau of Navigation (now personnel) established the Construction Organization as requested by BuDocks and authorized enlistment therein in this directive. By March 1, 1942, a small number of men had been recruited through voluntary enlistment and were being trained at the Naval Air Station, Quonset Point, RI, and at Youth Administration Camps, in the absence of a Seabee training camp.'

Since Admiral Wallin was the original commanding officer of the 1st Battalion, and perhaps quoting from his official records, certainly he must be considered an authoritative source.

At that time, I was a Chief Yeoman, V-6, USNR, reporting in from Washington, D.C., at the Naval Air Station, Quonset Point, on January 11, 1942. After a couple of days at NTC, Newport, for shots, pay records and medical check, I was advised that I would probably be sent to Iceland—a destination I was hardly enthused about.

At Quonset Point, I was the third Chief Yeoman on hand. The leading chief in the Exec's office sent me down to the station's Public Works Department where there was some story about a 'Mickey Mouse' construction detail that could use a Chief Yeoman 'to keep 'em straight.'

Orders from Washington to the C.O. of the air station assigned to me a number of men by name (an unusual practice then) directing, 'Enlisted Personnel: Please transfer the following named men to report to the Officer in Charge of Construction, Contract NOy 4175 for permanent duty in the Temporary Advanced Facilities Training School at the U.S. Naval Air Station, Quonset Point, R.I.'

Captain Raymond V. Miller, CEC, USN, headed the Public Works Department at Quonset Point. The first three officers who had just reported to him were Lt. Herbert M. Shilstone, Lt. (jg) T. J. Doyle (both assigned to develop some sort of a training program, and Lt. (jg) Charles Broadbent—all three were CEC Reserve officers. Lt. (jg) Broadbent had attended the Naval Academy and was released due to a football injury. He was made personnel officer, under whom I felt privileged to serve as leading chief at Quonset (and again, later, at the openings of both Camps Allen and Peary). All three were fine officers. Shilstone later was assigned to the staff of Admiral J. J. Manning, CEC, USN in the Atlantic. Broadbent went to the Pacific as Executive Officer of the 70th Battalion and Doyle's later service cannot be recalled.

At Quonset Point, an old wooden cafeteria building was used for offices and training. The Air Station was short on supplies, especially clothing, so a group of the first CPO's were sent one night to nearby Newport, RI, with a contract to purchase proper uniforms from a civilian outfitter.

An admittedly sketchy training program was put into operation. It was slanted especially at the erection of large, portable fuel tanks, welding, and the operation of mobile and portable construction equipment. It is believed that some of the men in the 'Bobcat' detachment did receive some of this training. It is understood that they had a rough time according to all reports. Among other foul-ups, the gear they received included heavy winter clothing, parkas and snowshoes, since they had originally been destined for Iceland.

As all now know, 'Seabees Can Do' and this first group out survived their vicissitudes and acquitted themselves nobly. The fuel tank farm which they built at Bora Bora supplied the ships and planes defending the sea route to Australia as well as those involved in the Battle of the Coral Sea.

As mentioned relative to the uniform supply, the Naval Air Station facilities at Quonset were most inadequate for the increasing stream of newly enlisted men. These were being minimally processed at the recruit station at Newport. Essentially, they were given medical examinations, shots, records and some clothing. Some 2,000 men had to be sent elsewhere, so they were dispatched along with newly-recruited or newly-recalled officers to the aforementioned Youth Administration camps throughout the eastern seaboard. With green officers and petty officers, it is a tribute to the men's maturity and experience, the manner in which all co-operated and pitched in to shape up into a semblance of organization.

The average age of the enlisted men was about 32, that of the officers approximately 28. Most, if not all, had some years of significant construction experience behind them, many having their own firms, or were construction superintendents, foremen and the like. One wonders what might have happened had they been the usual untrained Navy recruit of 18-20 years of age.

With the officers, an administrative snag had developed. Historically, the Navy had always considered CEC officers as 'Staff' only, never for 'Command' as such. With command authority necessary for the Battalion head, under the wartime pressure, the Navy conceded, solving the hitch, to enable recruitment of experienced construction supervisors, up to the rank of Lieutenant Commander, so that they might command the Battalions.

Somewhat the same applied to enlisted personnel, since, historically, they had been recruited as Apprentice Seamen, with advancement only after extensive training or experience and study. Since circumstances required skilled construction men, they tended to be older men with families. These men were typically draft-exempt, and in need of money for support of their families. By opening up the ratings to allow enlistment up through Warrant rank, it was thus made possible to procure a large number of patriotic volunteers who were skilled pro-fessionals in construction. There was some 'flack' from labor unions, but these were promised that, except for the battalion's own training facilities in the U.S., other work would be let out to contract by civilian labor. I have not been able to verify this information, but I have been told that perhaps 80% of all men in the wartime Seabees were union members.

There was initially some reaction from a few regular Navy enlisted personnel. The writer can recall occasional derogatory remarks made to CPO's about 'Slick-arm' chiefs, meaning those with no service stripes ('hashmark'). The average regular Navy CPO had two or three service stripes before making chief. The Navy tried, probably for similar reasons, to have special rating badges, with the letters 'CB' rather than the specialty symbol. The effort to have the men wear these failed, and after a couple of months the attempt was abandoned.

In order to provide a training facility, especially for the Construction Battalions, plans for one were rushed and ground was broken on January 17, 1942, at the Naval Operating Base in Norfolk. It was named Camp Allen, after Captain Walter H. Allen, CEC, USN, who had formed a construction regiment at the Great Lakes Naval Training Station in World War One. The N.C.T.C. at Camp Allen was supplemented by another, the first officer-in-charge of construction being Lt. Shilstone. Camp Bradford took its name from the farmer who sold his spinach fields to the Navy. This was located at the Little Creek

Amphibious Base, then known only as Little Creek, from the ferry terminal linking Norfolk with Cape Charles. This has now been supplanted with the Chesapeake Bay Bridge-Tunnel System.

Both of the Seabee facilities were temporary and had few amenities. Bradford, at first, was a muddy camp but improved as the war went on. The Amphibious Forces, which took over, expanded and continued to make improvements.

For recruit training, petty officers with some previous military training were selected and trained by experienced Marine drill instructors. As these became available, Camp Allen started to function as a recruit training center, and was formally commissioned on March 21, 1942.

In the past there has been some discussion on the exact date, but according to Admiral H. N. Wallin, CEC, USN, the official commissioning of the 1st Construction Battalion took place on March 15, 1942, at Norfolk, Virginia, only two days after the first men were re-assembled after their term in National Youth Administration camps.

Facilities at Camp Allen were hardly complete, some of the concrete floors in the two-story barracks buildings were still damp from their pouring. Catarrhal fever was prevalent, apparently encouraged by the never-ending dust coming off the concrete floors. I remember a desperate attempt to hold down the dust in the personnel office by using large quantities of 'water glass' (sodium silicate). I remembered this 'solution' from my youthful days of working for an egg dealer. We used water glass as a preservative for eggs and for package label adhesive.

I reported at Norfolk on March 8, 1942, where all was confusion. The men who had been at National Youth Administration camps reported shortly thereafter, Their arrival was complicated by the fact that one of the groups was almost entirely sick with the flu. Insofar as the personnel function was concerned, there was office talent galore among the new men, many having been businessmen, lawyers, accountants, insurance executives and the like. One of the men was a young, Boston lawyer, who had enlisted as a seaman 2c, with whom money was no object. He was a colorful character, remembered with great affection, as he was an invaluable assistant, a 'procurement' specialist of the finest talents.

As the nation 'revved up' to fight a global war, authorization followed authorization for more and more Seabees: 10,000, then 20,000, then 50,000, then 100,000, and so on. Y. H. Ketels, LCDR, CEC, USN (Ret.), Director, The Seabee Museum, Port Hueneme, California, states that there were a total of approximately 356,000 officers and men in the Seabees in WW II. Approximately 254,000 was the maximum on duty at one time. These figures are from the start of hostilities until the discharge system went into effect.

In addition, Y. H. Ketels states that while difficult to give the number of Seabees serving since WW II, the force was down to about 15,000 on duty during the Korean War, about 22,000 during the Vietnam War. During normal peacetime, the count remains approximately 10,000 male and female Seabees.

I recall being criticized by a ranking CEC officer in the early Camp Allen days, for his 'grandiose ideas for handling personnel paperwork for your mythical thousands.' Tabulating equipment was innovative then and I was trying to get authorization for several thousand dollars a month to rent keypunch and tab equipment to handle the vast detail of keeping records and forming battalions systematically. At that time some newly 'caught' yeomen, or others, were slaving 15-18 hours a day over a non-descript collection of used typewriters, making 3" x 5" cards to list the men and sort them into battalions.

One of my roommates, a Chief Storekeeper, caught a severe reprimand from a young, green Supply Officer for changing a signed order for 200 rolls of toilet paper to 200 cases. The same officer, seeing men unload cases of evaporated milk at the gallery entrance, inquired of the Chief, 'What is this, a week's supply?' To which the CPO had to reply that it was for the day. The average person inexperienced in such matters has no concept of the vast amount of supplies necessary for large numbers of men. My immediate superiors had been around from the start and were completely understanding. As a result, Camp Allen had one of the first three IBM personnel systems installed in the wartime Navy. The system was also duplicated at Camp Peary.

Gradually, things at Camp Allen became systematized, order prevailed, and things ran with comparative smoothness. Recruits were pouring in daily, receiving a fair semblance of military training, then turned over to their officers and organized as battalions.

An acquaintance, a junior grade warrant officer, who was a gunnery enthusiast later assigned to gunnery training, invented and developed a firing device for a mortar. This device utilized a blank .30 calibre shell, which fired a dummy mortar shell several hundred yards. One gradually came to the conclusion that these men *really could do anything*, and perhaps this is how the slogan 'CAN DO' came about.

According to the official instruction manual, each Battalion was to consist of a Headquarters Company of 173 men plus four companies, each with six platoons. These were: Platoon #1, 38 men, Maintenance; Platoon #2, 38 men, Construction; Platoon #3, 37 men, Construction; Platoon #4, 37 men, Blasting & Excavation; Platoon #5, 38 men, Waterfront (Longshoremen); Platoon #6, 38 men, Tanks, Steel and Pipe; the grand total being 1,077 men. The officer compliment of 32 comprised 10 CEC Warrant Officers, 2 Medical Officers, 2 Dental Officers, 2 Supply Officers, 1 Chaplain, and the rest Civil Engineering Corps. Of interest to many may be the base enlisted pay in 1942. Monthly: Chief Petty Officer, Permanent Appointment, $138; CPO Temporary Appointment (usually less than 1 year of service), $126; Petty Officer, First Class, $114; Petty Officer, Second Class, $96; Petty Officer, Third Class, $78; Seaman or Fireman, First Class, $66; Second Class, $54; Apprentice Seaman (Recruit), $50. The Seaman Recruit in 1983 received $551.40, or about four times a CPO's pay of WW II.

With the constant increase of manpower requirements, Camp Endicott, Davisville, RI (adjacent to NAS, Quonset Point) was commissioned August 11, 1942, and had a capacity of 11,000 men. Camp Peary, outside of Williamsburg, VA (a vast, wooded and swampy area notable for the muskrat population trapped for their fur) had a capacity of 40,000 men, and was commissioned in November, 1942. The writer also participated in the opening of this facility; serving until June, 1943. Later, another training center was opened at Camp Livermore, California.

Camp Peary is remembered as a large muddy mess. The rigors of living in the "boondocks" with little or no facilities could have been tolerated, except that we were all well aware that only a few miles away, life for the Williamsburg Restoration people and the university community seemed to be going on the same as before.

College football games were being played, crowds of fans were driving in from all over, in spite of gas rationing, and, most galling, especially for a man who had worked for a leading department store in Richmond in the 1930s, the village of Williamsburg, with all of its supposed historical significance, was declared off-limits to enlisted personnel. Undoubtedly, this was due to political pressure brought on by the restoration committee interests and/or the College of William & Mary. Originally a New Yorker, I spent most of my working life in Virginia, and my ancestors settled in the Virginia mountains in 1730. I never quite recovered from my feeling of exclusion from Williamsburg. The restrictions were later lifted, but it was most harmful to the morale of many patriotic Seabee volunteers who were gladly serving their country.

In the first days at Quonset Point, as the training program was being worked out, the camp was visited by Lt. W. B. Howard, CEC, USN, from the Bureau of Yards and Docks. He submitted the idea to the officers on board that it might be appropriate to find a shorter name than 'Construction Battalions.' In line with this, Lieutenants Shilstone and Doyle, the training officers, along with Lt. Broadbent in personnel, all became interested and solicited ideas. There were many. Some good ideas, some 'corny,' and none very appealing, such as 'pioneers,' or 'constructors,' or something in Latin. However, the officer's mess at that time was in a Quonset hut, and, on an off-duty Sunday evening, the matter was being discussed. A young Supply Corps reserve officer, Ensign R. D. Woodward, came up with the name 'SEABEE.' This was derived, of course, from the pronunciation of the first letters of Construction Battalions.

It was a natural and was immediately well received. The next step was to create a symbol. The thinking had been along the lines of a Walt Disney-type character, such as some of the aviation squadrons had started to use. Different ideas were tried, but, with the 'Seabee' name decided, the writer has always believed that the concept of the angry, fighting bee, with machine gun and tools, was definitely Lieutenant Shilstone's idea.

Eventually, the artwork was completed by Frank Iafrate, at the time a civilian plan file clerk in the Public Works Department. He later enlisted in the Seabees, but as a civilian, he utilized his talents in producing the first Seabee logo in full color. It should be noted that the 'Seabee' name is not in stencil type lettering and that the encircling border is the letter 'Q' for Quonset. The change to stencil lettering and the change of rope grommet for the 'Q' border came later. The changes were probably made in Washington before official acceptance of the name and symbol. The CEC officer symbol was also placed on the bee's rear leg.

By January, 1943, Camp Peary had several thousand men in the station crew (there were over 30,000 men in training). Literally hundreds were reporting in from recruiting stations daily. It was not unusual for us to process 500 to 600 men a day for records, physicals, shots, insurance, clothing, etc. A Battalion was shipped out every few days. It was an enormous undertaking, terrific pressure for all, the surprising thing being that the job was accomplished as well as it was.

This trip 'down memory lane,' as mentioned in the foreword, is intended primarily for my family history record. As historian for my local Navy Seabee Veteran's of America Island, I planned to make it available to any interested parties. Hopefully, it may be authentically informative.

Prior to writing this, most of the written material my possession, such as newspapers, clippings, early station orders from Camp Allen and Camp Peary, news bulletins, etc., which had been sent home during the war, were passed on to the Seabee Museum and Historian at the Port Hueneme, California base. Presumably, this sort of material can be made available to an interested researcher.

My narrative closes after the opening of Camp Peary. I was commissioned in January, 1943, and back-dated to November, 1942, as an Ensign, USNR, *but* as an aviation ground officer. My last official duties in personnel at Camp Peary were to coordinate and edit the first Battalion Administrative and Clerical Manual, plus help to organize a Yeoman school. I was then assigned as a Training Officer in small arms, and assorted other duties, before being transferred to Air Force, Atlantic Fleet. I experienced sincere regret at leaving the Seabees.

Regret? Because one can only say that overall the Seabees were the most capable, the friendliest, the most helpful shipmates any serviceman will ever know. Their 'CAN DO' spirit was real, invariably cheery, their confidence to accomplish always re-assuring, their ability outstanding and, above all, their superb fellowship was incomparable. End.

Reprinted, edited and condensed in the memory of Nathan S. Raitt , a former Life Member of SVA Island X-3, and with the encouragement of Richard A. Lindner, Commander, Dept. of New York, SVA, Island X-5, West Seneca, NY. Also, special thanks to Ann O'Day, Editor of CAN DO (716-684-1591) and Mel Ramige, National Secretary, SVA (1-800-SEABEE-5).

APPENDIX G

BRONX SEABEE FIGHTS FOR PROPER MONUMENT ON TINIAN AIR STRIP

by Joseph Garofalo (June, 1995)

I was a Seabee from November, 1942, to November, 1945. I was with the 121 N.C.B. that was formed at Camp Le Juene, North Carolina. We had been formed along with the 4th Marine Division and later went to Camp Pendleton, California for advanced training with that Division. Ultimately, we made 'D' Landings on Roi-Namur, Saipan and Tinian and were awarded the Presidential Unit Citation along with the 4th Marine Division.

We were the first Seabees on these invasions. Among the many memories there was the following incident that may be described as a classic now, but not then: On 'D' Day, June 15, 1944, we disembarked on an LCVT with approximately nine men. The rest of the cargo was ammunition. We had to go in on certain waves. Meanwhile, the landing crafts were to rendezvous until the unit is called to proceed toward the beach. We were given a mimeographed sketch of where to land on Blue Beach 2, to the right of the Japanese sugar mill on Chalon Kanoa, Saipan. All of a sudden, I couldn't see any of the LCVT's around us. Our coxswain was in total shock and heading for the island of Tinian—which was to be our next objective. Two divisions of Marines were heading for and landing on the island of Saipan. Our nine men were going to Tinian. This was not only the wrong island, but it had about 16,000 Japanese troops defending it.

One of the men on our landing craft slapped the coxswain in the face after I shouted, 'Wrong island!' After being slapped he regained his composure, turned around, and headed for Saipan. Meanwhile, the Japanese opened up on us. Two shells came so close that the amphibious craft almost turned over. Water came rushing in. Luckily, we made it back to Saipan, and that, of course, was no picnic either.

On Saipan I cut the leather belt that operated the sugar mill on Chalon Kanoa. It was 18" wide, eight stories high, or about 80 feet long. By coincidence, during my teens, I had helped an uncle in his cobbler shop in the Bronx. I had some knowledge of shoe repair, so, when time allowed, I cut slabs of this leather and repaired shoes and made knife sheaths for my buddies on Tinian. I ran the cobbler shop on Tinian with Lenny DeLunas of New Jersey and a Seabee from Georgia. I used the 4th Marine Division's machinery for making rubber heels and soles.

I visited Saipan and Tinian for the 50th anniversary of the invasion of Saipan, June, 1994. It was one of the highlights of my life and an experience I will never forget. The local residents were exceptionally friendly and cordial toward the returning veterans.

I spent one day on Tinian. I visited my old camp area which was about 100 yards from the Enola Gay and Bockscar loading pits. I was disappointed with the existing plaques honoring those two historic missions that ended the war in the Pacific and thus saved thousands of American and Japanese lives.

I witnessed a number of very substantial monuments that have been placed throughout Saipan and Tinian by the Japanese. Since returning home I have written over 100 letters to various organizations such as VFW, American Legion, Former Prisoners of War, Disabled Vets, Overseas Veteran, and the Seabee Veterans of America. Everyone agreed that a more suitable monument on Tinian was needed.

The Navy Department has rejected all Tinian peace ceremonies and monuments on Tinian's North Field as recommended by this writer. The difficult-to-accept rationale stated was ' To avoid the potentially volatile attention and adverse criticism that military support of such activities may cultivate.'

It was at North Field that I and approximately 14,000 other Seabees worked on the runways for the B-29 super fortresses that delivered the atomic bomb to Hiroshima and Nagasaki in August, 1945, thus saving lives and ending the war in the Pacific.

Persons interested in raising funds for a memorial monument on the North Field of Tinian can write to me, Joseph Garafalo, 1954 Hone Ave., Bronx, NY 10461

APPENDIX H

STAMP SERVICES

SEABEE STAMP PROPOSAL

February 21, 1997

Mr. Jay Kimmel
Publisher
CoryStevens Publishing, Inc.
640 N.E. 148th Ave.
Portland, OR 97230

Dear Mr. Kimmel:

Thank you for your recent letter to the Citizens' Stamp Advisory Committee expressing support for the issuance of a commemorative stamp honoring the Seabee Veterans of America/Naval Construction Forces of World War II.

I must inform you that the Citizens' Stamp Advisory Committee has previously reviewed the nomination of the Seabee Veterans of America/Naval Construction Forces, but it was not recommended for issuance.

Each year, we receive thousands of letters suggesting hundreds of different topics for new stamps. Since 1957, the Citizens' Stamp Advisory Committee has reviewed many worthy subjects and has recommended a limited number based on national interest, historical perspective, and other criteria. Unfortunately, a vast majority of suggestions submitted, including many meritorious and meaningful subjects, cannot result in a stamp.

As information, it is difficult if not impossible to individually honor the many elite units within the Armed Forces and their heroic members. Nevertheless, you will be pleased to learn that an authentic World War II vintage "Can Do" patch is featured on the cover of the 1945: *Victory at Last* book in the World War II series. This commemorative series is featured on page 13 of the enclosed *Stamps etc.* catalog.

We hope in some small way this acknowledgement pays tribute to the brave and hardworking officers and men of the Naval Construction Forces.

We appreciate your interest in our stamp program.

Sincerely,

James C. Tolbert, Jr.
Manager
Stamp Development

Enclosure

Navy Seabee Veterans of American (NSVA) and others have made a sustained effort to obtain approval of a U.S. postage stamp that commemorates U.S. Navy Seabees past and present. Obviously, continued, large-scale effort is required to obtain the Citizens' Stamp Advisory Committee's recommendation for issuance. SEABEES *Can Do.*

Sample of commemorative stamp proposed by NSVA, 1997

475 L'ENFANT PLAZA SW
WASHINGTON DC 20260-2435

BIBLIOGRAPHY

A Seabee's Diary of the Great Earthquake of 1989, Paz Gomez Allen, Navy Civil Engineer, Spring, 1990.

"A Stone's Throw from Tokyo: A Pictorial Overseas History of the 101st Seabees," 270 pp., 1945.

Birth of the Seabees [The], Vincent A. Transano, The Military Engineer, July 1, 1992.

"Build & Fight with the Seabees and Follow Your Trade in the Navy," 32 pp., Publisher: U.S. Navy, 1943.

Builders in Battle, Virginia Calkins, Cobblestone Magazine, January, 1994.

Building the Navy's Bases in World War II: History of the Bureau of Yards and Docks and the Civil Engineer Corps, 1940-1946, Vol. II, Publisher: Bureau of Yards and Docks, Government Printing Office, 1947.

Can Do! The Story of the Seabees, Wm. B. Huie, 250 pp., Publisher: E. P. Dutton & Co., 1944.

Can do! A national publication for all Seabees and CEC Corps, St. Louis: Navy Seabee Veterans of America, Inc., 1970.

"Can Do—Will Do!" A History of the U.S. Navy Construction Battalions, Jeffrey R. Millet, 160 pp., Publisher: Taylor Publishing Company, 1550 W. Mockingbird Lane, Dallas, TX 75235, 1987.

CEC/Seabee Historical Foundation, P.O. Box 657, Gulf Port, Mississippi, 39502, (601) 865-0480.

Combat 'Bees: "Fighting Seabees" Train, All Hands, August 1, 1990.

D-Day: The Seabees Were There, Palmer W. Roberts, Navy Civil Engineer, Summer, 1994.

Down Atabrine Alley with the 140th Seabees, 154 pp., Publisher: Army and Navy Pictorial Publishing Co., 1945.

Dr. Henry Story [The], U.S. Naval Facilities Engineering Command

End of an Era for the Seabees, Antarctic Journal of the United States, March 1, 1994.

Equipment Operator I & C, U.S. Bureau of Naval Personnel, Rev., 273 pp., 1966. Also 1971 edition (295 pp.) available through Supt. of Documents.

Fiftieth Seabees [The] 1945, 108 pp., San Francisco (?).

Fifty-fifth Seabees 1942-1945, Publisher: Army and Navy Pictorial Publg. Co., 1946.

Fifty-fifth Seabees [The] 1942-1945, Publisher: Army & Navy Publishing Co., 1945.

Fifty-fourth Seabees in the Philippines [The], Frank B. Rand, 1945 (?).

Fighting Seabees [The]: 50th Anniversary Edition, Publisher: Republic Pictures Home Video, 1993, 1994, [from original film starring John Wayne (1907-1979), and others).

From Omaha to Okinawa: The Story of the Seabees. William Huie, 1945.

"He Operated Under Viet Cong Fire," Ted Thackrey, Jr., Herald-Examiner Staff Writer, Herald-Examiner, Vol. XCVI, No. 73, June 7, 1966. Article about Lt. Harvey M. Henry of the Seabees (cf. "The Doctor Henry Story").

History of the Construction Corps of the U.S. Navy, Publisher: U.S. Navy Department, Bureau of Construction and Repair.

Laughing and Griping with the 97th Seabees, 189 pp., Publisher: Bay Port Press, National City, CA, 1983.

Long Ago & Far Away: The Story of the 63rd Seabees, Paul J. Matchuny.

Marvin Shields Story [The], U.S. Naval Facilities Engineering Command

Naval Facilities Engineering Command, 200 Stovall St., Hoffman II Building, Alexandria, VA 22332.

Navy Seabees [The]: Their Early Days, Nathan S. Raitt, 1983.

Northern Viking 93: The Seabees Were There, Daniel L. Grigsby, Navy Civil Engineer, Spring, 1994.

Oral History of the 27th Battalion in the South Pacific in World War II, Willard G. Triest, 1972.

"Out of Jungles Come Millions," James Joseph, Publisher: Nation's Business, October, 1951.

Pacific Album: The 75th Seabees, 196 pp., 1945.

Reserve Seabees Excel in the "Wooden Fist" Exercise, Bradley Posadas, Navy Civil Engineer, Winter, 1988.

Reserve Seabees Support NCEL, Bradley Posadas, Navy Civil Engineer, Spring, 1989.

Seabee Divers "Can Do," Publisher: All Hands, December 1, 1994.

"Seabee Reserve: Veterans Combine Training with Good Works," Publisher: Look Magazine, August 24, 1954.

Seabee: Bill Scott Builds and Fights for the Navy, Henry B. Lent, 176 pp., Publisher: The Macmillan Company, 1944.

"Seabees Build a Town" [Under Antarctic Ice], Richard F. Dempewolf, Popular Mechanics Staff Writer, Publisher: Popular Mechanics, Part One: April, 1956, Part Two: May, 1956.

Seabees Build Friendships in the Caribbean," Publisher: All Hands, December 1, 1994.

Seabees "Can Do!" : A Pictorial History of the Silver Anniversary," 274 pp., Publisher: Walsworth Publishing Co., 1967.

Seabees "Can Do!" A Pictorial History Commemorating the Silver Anniversary of the United States Construction Battalions, 274 pp., Pub: Walsworth Publishing Co., 1967.

Seabees Go to War [The], Karen Fedele, Navy Civil Engineer, Summer, 1991.

Seabees Help Communities in the Bahamas, Stephen T. Rookus, Navy Civil Engineer, Winter, 1993.

Seabees in Action, Publisher: Naval Facilities Engineering Command, 1966.

Seabees in Action: Story of the Seabees—World War II to Vietnam, LCDR William D. Middleton, Naval Facilities Engineering Command, 1966.

Seabees in Action: Vietnam Pictorial Report, Washington, DC: Naval Facilities Engineering Command, 1968.

Seabees in Profile: They've Got the Know-how, Publisher: All Hands, August 1, 1990.

Seabees in Somalia, Publisher: All Hands, February 1, 1993.

Seabees in War and Peace [The], Kimon Skordiles, Publisher: Argus Communications, 1973, 1976.

Seabees in World War II [The], Ben Moreell (1892-1978), 101 pp., Annapolis, 1962.

Seabees in World War II [The]: Thumbnail Sketches of Each Battalion, 23 pp., San Francisco, '83, '90. Micromaster *ZZ-1248.

Seabees Rebuild After Hurricane Andrew, George W. Powers, Jr., Navy Civil Engineer, Winter, 1993.

"Seabees Swarm Again [The]: The Navy's build-or-fight crews in Vietnam are barely out of their teens, but they're working the same kind of construction miracles that 'civilized'

advanced bases in World War II," Mort Schultz, Publisher: Popular Mechanics, March, 1968.

"*Seabees Time on the Air,*" (radio/TV program) Hy Freedman, 1945, 1963.

Seabees to the Rescue (Mt. Pinatubo), Navy Civil Engineer, Spring, 1992, & Summer, 1992.

Seabees, Marines Team Up to Improve Barracks at Camp, Rodney Gauthier, Publisher: Marines, May 1, 1994.

Seabees: The First and Finest, NY: U.S. Naval Mobile Construction Battalion No. 1, 1958.

Sixty-eighth Seabees, 59 pages, 1945.

Song of the Seabees [The], Lyrics by Sam M. Lewis, Music by Peter De Rose, Publisher: Robbins Music Corporation, © 1942 by U.S. Navy Bureau of Yards and Docks, Washington, D.C.

South Pacific Saga: 57th Seabees, 1942-1945, 182 pp., Publisher: Schwabacher-Frey Co., 1946, (cf. Sopac Saga).

Story of the 113th Seabees [The], 136 pp., Publisher: Schwabacher-Frey Co., 1947.

Super Breed I: A Journal of the First Marine Engineers, Pioneers and Seabees, 226 pp., Publisher: Wyatt Publishing, Tulsa, OK, 1990.

The Seabees in War and Peace, A Comprehensive Historical Account of the Activities of the U.S. Navy Men Who Helped to Build a Better World and Fought to Preserve It., Vol. One: To 1945, Kimon Skordiles, Publisher: Argus Communications, Inc., 2304 Huntington Dr, San Marino, CA 91108, 1973.

The Seabees of World War II, Edmund L. Castillo, 190 pp., Publisher: Random House, 1963.

There's a New Seabee Presence in the South Pacific, Gary A. Engle, Navy Civil Engineer, Spring, 1989.

Thirty-seventh Seabees, Publisher: Army & Navy Publishing Co., 66 pp., 1945.

Thirty-third U. S. Naval Construction Battalion, 85 pp., Army and Navy Pictorial Publishing Co., 1946.

Three Years with the Seabees, John W. Shaddix, East Texas State Teachers College thesis, 1947.

"*United States Navy Seabees, 1941-1945* (Audiovisual, b&w poster 91 x 61 cm), Publisher: National Archives and the Department of Defense, 1992. Subject: "Seabees repair airstrip on Tarawa with heavy grading equipment."

"*Viet Seabees Retrain,*" Ted Thackrey, Jr., Herald-Examiner Staff Writer, Herald-Examiner, Vol. XCVI, No. 70, June 4, 1966. Sub Head: "Heroes, hell, We're construction men!"

Wanted: Seabees for Underwater Construction Teams. USN recruitment brochure.

We Build, We Fight! The Story of the Seabees, Hugh B. Cave, 122 pp., Publisher: Harper & Bros., 1944.

We Build, We Fight: 50 Years of Navy Seabees (VHS video recording), Publisher: Delaney Communications, 1992.

"*We're Back in Business—the Seabees,*" Andrew Hamilton, Publisher: Popular Mechanics, May, 1951.

We've Got You Covered, 1948-1987, Alfred G. Don, National Historian, Navy Seabee Veterans of America, Inc.

With the Seabees in the South Pacific, Watt M. Cooper, 192 pp., Publisher: W. M. Cooper, Graham, NC, 1981.

"*Work-related Injury Frequency Rates in the Navy Seabees,*" 49 pp., James R. Van de Voorde, Seattle, 1991, (Univ. Wash. thesis).

World War II: The Story of the Formation of the U.S. Navy's Construction Corps—the Seabees. VHS.

NO CHALLENGE TOO BIG

Athey Wagon hauling coral rock (US Navy Photo, 1945)

INDEX